ORIGIN OF LIFE

ORIGIN OF LIFE

WHAT EVERYONE NEEDS TO KNOW®

DAVID W. DEAMER

OXFORD
UNIVERSITY PRESS

OXFORD
UNIVERSITY PRESS

Oxford University Press is a department of the University of Oxford. It furthers the University's objective of excellence in research, scholarship, and education by publishing worldwide. Oxford is a registered trade mark of Oxford University Press in the UK and certain other countries.

"What Everyone Needs to Know" is a registered trademark of Oxford University Press.

Published in the United States of America by Oxford University Press 198 Madison Avenue, New York, NY 10016, United States of America.

Library of Congress Cataloging-in-Publication Data
Names: Deamer, David W., 1939– author.
Title: Origin of life : what everyone needs to know / by David Deamer.
Description: New York, NY : Oxford University Press, [2020] |
Series: What everyone needs to know |
Includes bibliographical references and index.
Identifiers: LCCN 2020003142 (print) | LCCN 2020003143 (ebook) |
ISBN 9780190098995 (hardback) | ISBN 9780190099008 (paperback) |
ISBN 9780190099022 (epub) | ISBN 9780190099015 (updf)
Subjects: LCSH: Life—Origin.
Classification: LCC QH325 .D425 2019 (print) |
LCC QH325 (ebook) | DDC 576.8—dc23
LC record available at https://lccn.loc.gov/2020003142
LC ebook record available at https://lccn.loc.gov/2020003143

CONTENTS

SECTION 3: WHAT WE STILL NEED TO DISCOVER   61

INTRODUCTION

I'll begin with a challenging question: Why should anyone want to know about the origin of life? The answers will vary from one person to the next, but the simplest answer is curiosity. Anyone reading this introduction is curious because they wonder how life could have begun on the Earth, but there is more to it than that. My friend Stuart Kauffman wrote a book titled *At Home in the Universe*. The title refers to a deep sense of satisfaction that comes when we begin to understand how our lives on Earth are connected to the rest of the universe. There are surprises and revelations as we discover those connections. For instance, living cells are mostly composed of just six elements. As you read further in the book, you discover that the hydrogen atoms in your body are 13.8 billion years old, as old as the universe, and the rest of the atoms were synthesized in stars over five billion years ago. Life on Earth borrows those atoms from the universe for a brief time and then gives them back.

Then there is a practical consideration. Curiosity-driven research can satisfy scientific curiosity, but the discoveries we make sometimes have valuable spinoffs. My own experience in this regard was an attempt to make a laboratory model of a primitive cell. I needed to find a way to get a molecule called adenosine triphosphate (ATP) through membranes so that enzymes encapsulated in lipid vesicles would have a source of energy to synthesize ribonucleic acid (RNA). Asking that question and finding an answer a few years later led to a method called nanopore sequencing of deoxyribonucleic acid (DNA), and commercial instruments that incorporate the original idea are now being manufactured. The cycle of discovery and invention is now coming full circle back to basic research because nanopore devices are being developed that will search for life elsewhere in the solar system.

Astrobiology helps us understand how life can begin

The first speculations about how life can begin on Earth were published in a Russian language book by Alexander Oparin in 1924, followed by J. B. S. Haldane's brief essay in 1929. Both concluded that life's origin can be understood in terms of chemistry, and this view has guided research ever since. However, a newly emerged discipline called astrobiology expanded our perspective beyond the Earth and its biosphere. Astrobiology builds on ever-increasing knowledge of how planets, stars, galaxies, and even the universe began. We now have a pretty good idea how the Earth became a habitable planet and why life is likely to be distributed throughout our galaxy with its 150 billion stars and planets, some of them surely habitable.

There are many pieces to the puzzle of how life can begin, and many ways to put them together into a "big picture." Some of the pieces are firmly established by the laws of chemistry and physics. Others are best guesses about what the Earth was like four billion years ago, based on reasonable extrapolations of what we know by observing today's Earth and other planets in the solar system. There are still major gaps in our knowledge, and this is where scientists can have radically different opinions about plausibility. For instance, which came first: Metabolism or genes? Proteins or nucleic acids? Most will agree that liquid water was necessary, but was it in a hydrothermal submarine setting in seawater or a freshwater site associated with emerging land masses? Let's first consider how scientists working on such problems try to find answers.

What is the difference between an idea, a conjecture, a hypothesis, and a theory?

Everyone has ideas, and there's an old saying that ideas are a dime a dozen. The reason ideas are so common is that people generally enjoy pondering questions and coming up with possible answers. A conjecture is a fancy name for a complicated idea that attempts to explain something specific. Even though a conjecture might sound reasonable, it probably does not have a solid foundation of facts. In his book *Life on the Mississippi* Mark Twain wrote, "There is something fascinating about science. One gets such wholesale returns of conjecture out of such a trifling investment of fact." Twain's insight strikes close to home when it comes to the origin of life: few facts and lots of conjecture. On the other hand, Twain was a great writer but he was not a scientist. During his lifetime, a few pioneering scientists were beginning to explore physics and chemistry using a tool called the scientific method.

What is the scientific method?

In high school, most of us learn about the scientific method, typically defined as a process involving five steps: 1. Make observations. 2. Perceive an interesting question. 3. Propose a hypothesis. 4. Test the hypothesis experimentally or by further observations. 5. On the basis of the positive or negative results, decide whether the hypothesis is correct, or at least has explanatory power. If the explanation is significant, repeatable by others and reaches a consensus, the hypothesis becomes a theory.

This sounds like a reasonable way to understand the world we live in, but in real life the process is quite a bit messier, at least for origins of life research. There is so much we don't understand that each researcher has only a vague idea of the big picture, and their ideas often contradict the ideas of other researchers. What we do know with certainty is that the origin of life occurred within the framework of the immutable laws of physics and chemistry, so the goal of science is to use those laws to fill in the vast gaps in our knowledge and perhaps someday to understand how life can begin.

Can life be defined?

There is little agreement about a dictionary-style definition of life that can be stated in one sentence. The reason is that cells, the units of life, are not things, but instead are systems of molecular structures and processes, each of which is necessary for the function of the whole. However, it is possible to list the most general properties and then describe the individual structures and processes in such a way that when taken together they can only fit something that is alive. Maybe that's the best we can do, so here are some general properties followed by a list of twelve specific properties that define cellular life on Earth.

General properties

Living cells are encapsulated systems of polymers that use nutrients and energy from the environment to carry out the following functions:

Enzyme-catalyzed metabolism
Growth by catalyzed polymerization
Guidance of growth by genetic information
Reproduction of genetic information
Division into daughter cells
Mutation
Evolution

Specific properties

1. A living cell consists of two basic kinds of polymers encapsulated by a membranous boundary. The polymers are composed of six elements abbreviated CHONPS, for carbon, hydrogen, oxygen, nitrogen, phosphorus, and sulfur.

2. One kind of polymer includes proteins that can either be structural or able to function as enzyme catalysts. The other kind of polymer is called nucleic acid and contains genetic information in the sequence of its monomers.

3. The monomers of proteins are twenty different amino acids, and the monomers of nucleic acids comprise eight different nucleotides, four of which compose DNA and the other four RNA.

4. Living cells require a source of nutrients from the external environment.

5. Living cells require a source of energy such as light or the chemical energy in nutrients. The energy is used to drive metabolic reactions that change nutrients into the compounds used by life.

6. Polymerization does not occur spontaneously, but requires a source of energy. As a result of metabolism, the monomers of proteins and nucleic acids have chemical energy added to their structures which allows enzymes to link them together into polymers.

7. Enzymes catalyze the synthesis of proteins and nucleic acids, and the process is guided by genetic information in the nucleic acid polymers. Proteins are synthesized by intracellular structures called ribosomes.

8. As a result of polymerization reactions, cells grow and duplicate the polymers containing genetic information.

9. The nucleic acid called DNA can be replicated in a process catalyzed by the enzymes.

10. At a certain point in the growth process, cells with duplicated genetic information divide and thereby reproduce.

11. Errors called mutations occur in the duplication process, so that individual cells in populations such as a bacterial culture have variations in their genomes.

12. Some of the variations provide a selective advantage, and those cells and their progeny survive, while those lacking the advantage are left behind. This process is called evolution.

These are the properties of living cells, and it is obvious that they are components of an incredibly complex system. When we try to relate these properties to the origin of cellular life, it is helpful to consider them one at a time as a series of questions.

- Where did the membranes come from that were required to form the boundaries of the first cells?
- What energy source was used by the first cells?
- What organic compounds were available, and where did they come from?
- How did metabolism begin?
- How did life become homochiral?
- What were the first polymers related to life?
- How were those polymers synthesized before life began?
- How were polymers encapsulated within membranous boundaries?
- How did certain polymers become catalysts?
- How did other polymers begin to contain genetic information?
- How were those polymers able to grow and replicate?
- What were the first forms of nucleic acids?
- What were the first forms of proteins?
- How did base sequences in nucleic acids begin to guide amino acid sequences in proteins?
- How did cells begin to divide and reproduce?
- What were the first steps of evolution?

These questions represent the edge of what we know about the origin of life and will be used to guide the nuggets of knowledge presented in this book. The nuggets have been discovered by a few researchers who are brave enough to venture beyond the edge of the known into the unknown. They lack maps but are sustained by the knowledge that life did have a beginning, and that even extremely unlikely processes become virtual certainties given a hundred million years and the immense surface area of a habitable planet like the early Earth.

Most people think of science in terms of answers that can be read about in textbooks, but working scientists know better. They know that even though answers are the valuable output of science, the excitement lies in the unanswered questions they spend their lives addressing. This book is organized in three sections that reflect both the answers and questions. The first section, "How to assemble a habitable planet," traces what we know about the history of the biogenic elements from their origin in stars to their delivery to the Earth and other habitable planets in our galaxy. The second section, "From not alive to almost alive," describes how simple organic molecules became increasingly complex over time, eventually assembling into structures that are almost, but not quite, alive. The third section, "What we still need to discover," addresses the questions still to be answered if we are to understand how molecular structures that are almost alive become alive. Although we will never know with certainty how life did begin, it does seem possible that we will understand how life can begin on any habitable planet such as the early Earth.

Section 1

HOW TO ASSEMBLE A HABITABLE PLANET

Hydrogen is a colorless, odorless gas that, when given enough time, changes into people. How much time? 13.7 billion years!

A good way to present the information in this book is to pose a question, provide an answer, and then go on to ask and answer another question: How do we know? The first question is obvious: Can hydrogen really change into people? To answer that, we need to start with a simpler question: Where do the atoms of life come from?

The elements of life on Earth are billions of years old

It's astonishing to realize that the carbon, oxygen, and nitrogen atoms in water and in proteins, nucleic acids, and cell membranes are billions of years old. In fact, the number of hydrogen atoms in a human body is roughly 70 percent of the total number of its atoms, and hydrogen atoms are 13.7 billion years old, as old as the universe itself.

Can this possibly be true? Keep in mind that science does not deal in certainty, but instead proposes explanations that best fit the evidence and then tests the explanations by further experiments and observations. For instance, seventy years ago there were two alternative explanations of the origin of the universe. One was a steady state theory promoted by Fred

Hoyle that said the universe did *not* have a beginning. The second, proposed by George Gamow, was that the universe *did* have a beginning—the term "Big Bang" was coined by Hoyle to describe this idea of a beginning. Gamow's theory made two important predictions: the universe should be expanding, and there must be a rumbling radio frequency left over from the Big Bang, something like the echoing thunder that follows a lightning strike.

How do we know?

Observations by astronomers showed that the light from galaxies millions and even billions of light years away from Earth was shifted from blue toward red, from shorter to longer wavelengths, which was consistent with a universe that was expanding from a point in time 13.7 billion years ago. Radio astronomers also observed a constant static in the microwave frequency range that seemed to come from all directions. This was predicted by Gamow and is now called the cosmic background radiation. Hoyle's steady state idea was abandoned, and the Big Bang theory has been accepted by consensus because it has more explanatory power.

Plate 1 shows what the universe looks like today. Most of the visible matter in the universe has gathered into galaxies like our Milky Way. The galaxies are not dispersed at random but instead form the clusters that are visible in the image. What you see are billions of galaxies, each with billions of stars, all powered by energy released when the protons of hydrogen fuse to form helium. (At the temperature of stars, a hydrogen atom cannot keep its electron, so only protons undergo fusion reactions.) The clusters occur because hydrogen, like all matter, has gravity, and gravitational force can cause hydrogen to gather first into clouds of gas, then into slowly spinning disks, and finally collapse into a star at the center of the disk, often with planets circling around it. Gravity holds the stars together in galaxies, and gravity causes galaxies to form the clusters seen in the image, which are not imaginary. This is an actual map of galactic clusters based on observational results. The central white line from left to right is not an artifact. It is the Milky Way, our own galaxy, which is viewed edge on so that it hides the galaxies behind it. The dark areas obscuring some of the starlight is interstellar dust that is ejected from dying stars and then gathers into vast clouds.

Atoms heavier than hydrogen are synthesized in stars

If stars did nothing but fuse hydrogen into helium, the universe would be lifeless. However, nuclear chemistry occurring in stars includes a second fusion process in which the elements of life—carbon, oxygen, nitrogen, phosphorus, and sulfur—are synthesized, along with the iron and silicon composing rocky planets like the Earth. When an ordinary star runs out of hydrogen fusion energy, it first expands into a red giant and then collapses and releases most of its remaining mass as microscopic particles called interstellar dust. These particles are composed of silicate minerals and iron mixed with water and organic compounds containing the biogenic elements, together with trace amounts of other elements in the periodic table.

How do we know?

As ever-increasing technological knowledge has allowed us to construct powerful telescopes, we can actually see elements being ejected from stars that collapse when they "run out of gas" and no longer have a source of fusion energy. These telescopes aren't necessarily the sort that have glass lenses and mirrors to collect visible light. Radio telescopes can "see" radio waves; infrared telescopes can confirm the existence of organic molecules in space; and other telescopes can build up images from ultraviolet light and X-rays. There are even telescopes like the Hubble in orbit circling the Earth, far above the atmosphere that distorts the light from distant stars and galaxies.

The image shown in Plate 2 is the remains of a supernova as seen by an X-ray telescope. It is called Cassiopeia A, and the image has been color coded to show which elements have been ejected from the collapsing star. The purple color is iron, yellow is sulfur, green is calcium, and red indicates silicon. All that's left of the star itself is the tiny dot in the center called a neutron star.

Iron, sulfur, and calcium are used by life processes. But what about carbon, oxygen, and nitrogen? Where do they come from? In the late 1940s the British cosmologist Fred Hoyle had an idea. He could account for the synthesis of carbon in stars at sufficiently high temperatures if beryllium, with four protons in its nucleus, fused with an alpha particle composed of a helium nucleus. Oxygen and nitrogen can then be synthesized from the carbon by the carbon-nitrogen-oxygen (CNO) cycle shown in Plate 3. In the star we call our sun, characteristic wavelengths of the various elements can be seen in the spectrum of sunlight. With the exception of hydrogen and helium, the amounts of the elements present in the sun are similar to the amounts in the surrounding planets, a clear indication that the entire solar system was formed within a vast molecular cloud of dust and gas.

Six biogenic elements compose all forms of life

The biogenic elements are simply those that compose most of the mass of a living organism. Because a typical living cell is around 60–70 percent H_2O, or water, the major fraction by weight is the oxygen in the water. Carbon is the next most abundant element because it is present in all biomolecules like proteins and nucleic acids, along with nitrogen. However, in terms of the number of atoms in a living cell, hydrogen is the most abundant and represents approximately 70 percent of the atoms.

How do we know?

Suppose we take one gram of living bacteria, such as those that cause milk to become sour or apple cider to turn into vinegar. The bacteria are certainly alive, and the acids they produce—lactic acid and acetic acid—are the waste products of their metabolism. Next we heat the bacteria to 600 °C in a vacuum, a process called pyrolysis, which causes all the organic molecules to break down into atomic elements and a few simple molecules like water. Pyrolysis turns the bacteria into black ash, and when the ash is analyzed it turns out to be mostly composed of elemental carbon mixed with small amounts of chloride salts of sodium, potassium, magnesium, and calcium along with a little phosphate and sulfur. When we analyze the gas that is given off, it is mostly water—H_2O—with smaller amounts of nitrogen gas and sulfur as hydrogen sulfide, or H_2S. Finally we weigh the black carbon and calculate the mass of material in the gases.

The results, after a bit of arithmetic, can be illustrated either as the percent by mass or by numbers of atoms. Oxygen composes most of the mass of a living cell because it is in water (H_2O). In terms of the numbers of atoms, hydrogen is 62 percent of the total mass, followed by oxygen, carbon, and nitrogen in proteins and nucleic acids (Figure 1.1). Although phosphorus and sulfur are essential for life, they compose just

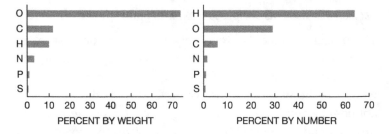

Figure 1.1 Biogenic elements in a bacterial cell. O, oxygen; C, carbon; H, hydrogen; N, nitrogen; P, phosphorus; S, sulfur.
Source: Author

a tiny fraction of the elements. The thing to keep in mind is that all of the biogenic elements except hydrogen and a little helium were synthesized in stars by stellar nucleosynthesis. The hydrogen in the water, proteins, and nucleic acids of life is there only because it did not get caught up in star formation. This is why a major fraction of the elements of life is as old as the universe.

Interstellar dust provided the atomic and molecular seeds of life for the solar system

The image shown in Plate 4 was taken by the Hubble telescope orbiting the Earth. It shows a beautiful spiral galaxy called NGC 1566, which is an abbreviation used in astronomy for New General Catalog, followed by a number. The galaxy contains billions of stars, and regions of new star formation in the arms can be recognized by their reddish glow. But something else is also visible, the dark bands that accumulate within the galaxy and hide the light of stars. We see similar bands in the Milky Way, our own galaxy. Those dark regions are called interstellar dust, composed of the ashes of stars that have gone through their lifetime and then exploded. The stardust contains the elements that were synthesized within the stars, such as iron particles and silicon in the form of silicate minerals. A thin film of water ice accumulates on the surface of the dust particles, and the ice contains simple molecules such as carbon dioxide (CO_2), carbon monoxide (CO), ammonia (NH_3), and methanol (CH_3OH).

How do we know?

In 1932, Karl Jansky was working at Bell Telephone Laboratories and trying to figure out sources of noisy static that were causing problems for radio communication between countries. He noticed that the static grew louder once a day when his antenna was pointing at the center of the Milky Way.

It was hard to believe, but stars were broadcasting radio waves. Over the next fifty years the antennas and amplifiers were refined to the point where it became possible to detect not just the radio waves but also modulation within the waves caused by organic molecules. Over a hundred such compounds are now known, and quite a few are related to the origin of life. These are composed of the biogenic elements, including water (H_2O), carbon dioxide (CO_2), hydrogen cyanide (HCN), ammonia (NH_3), formaldehyde (HCHO), formic acid (HCOOH), acetic acid (CH_3COOH), and even an amino acid called glycine (CH_2NH_2COOH).

Later research led to the conclusion that ultraviolet light causes simple molecules in the ice that coats interstellar dust particles to form more complex organic compounds. These were transported first to the solar system and then to the Earth late in the planet-forming process, after it was cool enough for an ocean to form. Other essential compounds were synthesized at the Earth's surface by chemical reactions in the atmosphere, ocean, and volcanic land masses.

Molecular clouds are the birthplace of stars and planets

The image in Plate 5 shows molecular clouds in a star system called Rho Ophiuchi, about 460 light years distant from Earth. A light year is simply the distance that light travels in a year, equivalent to 5.9 trillion miles. Our galaxy, the Milky Way, is at least 170,000 light years in diameter. To give a sense of how big this really is, the star nearest the Earth is called Alpha Centauri and is 4 light years away, and molecular clouds range from 3 to 70 light years in size. In comparison, our solar system is minuscule, just 4.2 light hours from the sun to Neptune!

In a very real sense, molecular clouds are the ashes of long-dead stars that have gone through their lifetime, and then exploded and ejected their elements into interstellar space. Some of the microscopic dust particles in the clouds are composed of silicate minerals, while others are metallic iron and nickel.

The dust particles in this image glow blue because they are reflecting light from nearby stars, while others range from brown to black. Elsewhere in the cloud, ultraviolet light causes hydrogen gas to glow red, an emission something like the red glow of neon caused by electrons moving through the gas. Molecular clouds are an important component of our understanding of life on Earth because they are the birthplace of new stars and solar systems.

How do we know?

The answer is simple. The Hubble telescope is in orbit 353 miles above the Earth's atmosphere, and it has provided incredibly clear views of processes occurring in the molecular clouds of dust and gas scattered throughout our galaxy. With the Hubble telescope we can look deep into nearby molecular clouds and see new stars forming everywhere, some surrounded by the dust that will form planets.

The solar system assembled from a disk of dust and gas circling the sun

A new star rises, like a Phoenix, from the ashes of dead stars that have been cremated by the final heat of 100 million degrees before collapsing and then exploding as a nova or supernova. Because new stars emerge in a molecular cloud, they usually appear in clusters such as the Pleiades constellation visible in the night sky. Their radiation drives away the dust of the cloud, and after millions of years only the cluster remains. The sun was once a member of such a cluster, but over the five billion years since it became a star its sister stars slowly moved away into interstellar space. The interstellar dust particles and gas gather into clouds from which new stars emerge and turn into rotating disks. Planets form as the dust in the disk undergoes gravitational accretion into kilometer-sized planetesimals, which then collide to form planets. The asteroids in orbit between Mars and Jupiter are planetesimals that did not get caught up in planet formation.

How do we know?

Plate 6 shows a dust particle and a molecular cloud in the background in which gravitation has begun to cause stars to form from the dust and gas composing the cloud. A theoretical explanation of planet formation was presented many years ago, but there was no direct evidence. However, a new telescope in Chile called the Atacama Large Millimeter Array (ALMA) can actually see what appears to be a developing solar system in a nearby star called HLTauri (Plate 7). The star is only a million years old and is surrounded by a disk of gas and dust, just as predicted by theory. The obvious gaps in the disk are most likely being produced by new planets as they gather up the dust. It is reasonable to assume that our solar system was formed by a similar process.

Radioactive elements keep the Earth's core molten

Later in the book we will describe how volcanoes emerged through a global ocean and formed the first dry land. Water that was distilled from the salty ocean by evaporation fell on the volcanic islands as rain. Volcanoes are hot, so the rain resulted in hot springs resembling what we see today in volcanic regions around the world. But why were there volcanoes? And how can they still exist today, four billion years after the origin of life?

The origin of the Earth involved gravitational accretion of asteroid-sized bodies called planetesimals that were many kilometers in diameter. The energy released by the impacts was so great that the iron and silicate minerals became molten as the Earth grew in size. Toward the end of the primary accretion, the orbit of a Mars-sized object happened to cross the orbit of the Earth, and the two planets collided and merged.

The moon formed from the hot silicate mineral debris that went into orbit around the Earth. Both the Earth and moon were heated by the impact to the temperature of molten lava. Plate 8 shows an artist's rendition of what our planet looked like at the time.

Because the entire Earth was molten, the dense iron and nickel that had been delivered during accretion sank through the lighter silicate minerals of the crust to form the core, which then began to cool. The diameter of the core is around 20 percent that of the Earth, and its temperature is estimated to be 6000°C, which is as hot as the surface of the sun! That heat is what powered volcanoes on the early Earth, and still powers them today. But there is a problem. When we measure the rate at which heat is being lost through the crust to outer space, the primordial heat of accretion could not have maintained the core at this temperature. There must be another source of heat.

How do we know?

The answer comes from our knowledge of long-lived radioactive isotopes mixed in with the iron core. The element with the longest half-life of 14.1 billion years is thorium-232, followed by uranium-238 (4.47 billion years), potassium-40 (1.28 billion years), and uranium-235 (705 million years). The number behind the name of the element is the atomic weight of the particular isotope of the element, which is basically the combined mass of protons and neutrons in the nucleus. The half-life is the amount of time it takes for half of the element to undergo radioactive decay into other elements with the release of heat energy. After four billion years, most of the original potassium-40 and uranium-235 would have decayed, so today the heat is being generated by thorium-232 and uranium-238.

About a billion years ago, the core had cooled sufficiently to separate into a solid inner core of iron and a fluid outer core. The slow convection of the molten iron of the outer core is the reason that the Earth has a magnetic field. This is significant, because the magnetic field deflects much of the potentially dangerous high-energy solar wind emitted by the sun.

Later I will argue that life could not have begun unless there were volcanic islands with a source of freshwater from precipitation. In other words, the fact that living organisms are so abundant on Earth today depends on a core of iron with a temperature sufficient to keep it in a fluid state. It is significant that Mars once had shallow seas and volcanoes. I would gladly make a $100 bet that we will discover evidence of life existing on Mars 3.5 billion years ago. Remnants may still thrive deep underground where residual heat keeps a little water in a melted state.

Radioactive decay tells us the age of the Earth

Our planet is approximately 4.57 billion years old, which is one third the age of the universe.

How do we know?

The age of the Earth has been determined in several ways. The simplest to understand is based on the fact that uranium is radioactive and decays into lead at a certain rate. For instance, suppose that we had a pure sample of an isotope of uranium called $^{238}U_{92}$ and measured changes in its radioactivity over time. By extrapolation, the result showed that half of it would turn into lead (abbreviated $^{206}Pb_{82}$) in 4.468 billion years. This is called its half-life. The second number in the abbreviation is the number of protons in the atom, which is fixed, and the first is the atomic weight that includes the number of protons and neutrons. Different isotopes of an element have different atomic weights. For instance, $^{235}U_{92}$ is the explosive isotope of uranium used in nuclear reactors. It has three fewer neutrons than $^{238}U_{92}$ and a shorter half-life of about 704 million years.

The next step is to assume that the uranium found in an ancient crystal of the mineral zircon was pure to begin with. We know that it was pure because lead does not fit into the crystal lattice of the zirconium oxide mineral, but uranium does. When we measure the amount of the two isotopes in the oldest zircons, it turns out that the ratio is close to 1:1, which means that half the uranium has decayed into lead and the zircon must be 4.5 billion years old. When we make the same measurement with the uranium and lead in a meteorite, the ratio is also 1:1. Finally, assuming that meteorites formed in the solar system at about the same time as the planets, the most careful measurements indicate that their age is 4.57 billion years, so we take that as the age of the Earth.

It's interesting to get a sense of how far back in time this actually was. Imagine that you are sitting in a time machine, a device that can transport you far into the past. You set the dial to four billion years ago, and the rate at 1000 years per second, and then push the button marked GO. Looking out the window five seconds later you see pyramids being built in Egypt and in ten seconds tribes are planting crops along the River Euphrates in the Middle East. Thirty seconds back in time you watch artists painting wild cattle on cave walls in France. At the seventy-second mark tribes are leaving Africa and wandering up into Europe, and at three minutes the first human beings, *Homo sapiens*, appeared in Africa.

After that, I'm afraid you need to sit in the time machine for eighteen hours until there is an enormous flash of light followed by a brief period of complete darkness caused by an asteroid six miles in diameter that crashed into the sea near Yucatan in Central America, causing the extinction of the dinosaurs that had ruled the Earth for 200 million years. Some small, warm-blooded mammals survived, or we wouldn't be here.

And now, more waiting. In six and a half days you come to the Cambrian, when some of the first animals dominated the ocean and became fossilized as trilobites. The land was also just beginning to turn green as plants learned how to live out of water by using sunlight as an energy source.

More waiting. This is getting boring! After twenty-nine days you are getting short of breath because there is almost no oxygen in the atmosphere. The main form of life is microbial, and you can see the ocean colored green from all the cyanobacteria thriving in it. They are making oxygen but not yet enough to breathe, so you put on an oxygen mask.

Finally, forty-six days later, you look out the window and can only see an ocean and volcanoes. There is nothing alive, so you have come to the prebiotic Earth. When you look at the volcanoes more closely you can see hot springs and geysers making small puddles of water with frothy bubbles around the edges. When they dry out a thin film is left, something like bathtub rings. When rain again fills the puddles, the compounds in the rings disperse into the water as microscopic vesicles. They are not alive, but they are the first step toward cellular life. Just give them a little time, for instance 100 million years, and they will manage to find a way.

Life could not begin until there was an ocean

The Earth at the time of life's beginning had a salty ocean with volcanic land masses emerging into an atmosphere mostly composed of nitrogen and a small amount of carbon dioxide. Because the Earth was still cooling from its molten state, the global temperature was much warmer than today, probably in the range of 60 to 80°C. There were no continents because the process of plate tectonics had not yet begun, but there were volcanic island land masses resembling Hawaii and Iceland. Precipitation produced freshwater pools on volcanic islands that were heated to boiling by geothermal energy and then cooled to the ambient temperature by runoff. Contemporary examples include the hydrothermal fields of Kamchatka, Hawaii, Iceland, and New Zealand.

How do we know?

There are three ways to learn what the early Earth was like. The first is from our knowledge of geology and mineralogy. Because of plate tectonics, there is almost nothing left of the original Earth's surface except a tiny fraction of rocky land in northeastern Canada that has been dated to 4.03 billion years old. This is approximately the time that life began, so we know that an ocean must have been present. We also know from the composition of ancient sedimentary minerals that there was no oxygen in the atmosphere.

The second method depends on the atomic composition of zirconium minerals called zircons that were isolated from sedimentary rocks in Western Australia and dated to 4.4 billion years old. The zircon composition can be used to estimate the temperature at which the rocks formed. The temperature was surprisingly low, well below the temperature of molten lava, which means that liquid water in the form of an ocean was present.

The last method is simply to examine the moon, which is covered with enormous numbers of craters resulting from impacts of large and small asteroid-sized objects. This is referred to as the Late Heavy Bombardment (LHB), which ended around 3.8 billion years ago. This means that the Earth when life began was a dangerous place. In fact, it has even been suggested that life could have begun several times but was wiped out by the violent impacts.

Life must have emerged sometime between 4.57 and 3.46 billion years ago, which are the dates established for the age of the Earth and the age of the first fossil evidence for microbial life. We can refine the time by assuming that life could not begin until liquid water was available, and we know from zircons and geological evidence that a global ocean was present about 4.3 billion years ago. There is also evidence that the Earth was being bombarded by asteroid-sized objects early in its history. Vast lunar craters, like Imbrium, are the scars left by such impacts. Because Earth is so much larger than the moon, it would have had even greater numbers of impacts. The energy delivered by the largest impactors could have caused extinction of any primitive life that had got started. It has been suggested that life may have begun on several occasions and only survived after the bombardment ended around 3.8 billion years ago. From these considerations, a reasonable guess is that life began sometime between 4.2 and 3.8 billion years ago.

Earth's water was delivered by planetesimals and comets

The dust in molecular clouds has a thin layer of ice coating the silicate mineral matrix. In the early solar system, the dust gathered by gravitational accretion into planetesimals and comets, along with the water, which was then delivered to planets like the Earth and Mars during their formation. The Earth was very hot at first, and an entire ocean of water was present as vapor in the atmosphere. At some point, the Earth cooled to a temperature at which water could exist as a liquid to form a global ocean. The early atmosphere was composed mostly of nitrogen gas with a little carbon dioxide. Because photosynthesis had not yet begun, there was no free oxygen.

How do we know?

Although water is a clear liquid when viewed by ordinary light, it would look opaque if you viewed it using infrared light because water molecules absorb the photons and turn them into heat energy. Water also absorbs other forms of electromagnetic energy. For example, the microwaves used to heat food and water in a microwave oven have a frequency of 2.45 billion per second, and their vibrational energy is absorbed by water, which then heats up. This is similar to the way your skin gets warm when exposed to the infrared radiation of sunlight. Radio astronomers aim their telescopes at molecular clouds, and the microwave spectrum provides clear evidence of water that is present as icy films on the surfaces of the dust particles.

During planet formation, the dust forms clumps by gravitational attraction, bringing along the water and organic compounds. The clumps grow until they become planetesimals the size of asteroids and comets, which range from tens to hundreds of kilometers in diameter. In the inner solar system close to the sun, the dust and water are pushed out past the orbit of Mars and then form the giant planets of Jupiter, Saturn, Uranus, and Neptune. Planets in the inner solar system result from collisions of planetesimals and comets and are rocky and much smaller. The Earth was able to keep a little of the water that arrived on planetesimals and comets and later condensed into the ocean. Mars had even less water, most of which evaporated into space during the past three billion years.

Because the ocean seems so large, covering two thirds of the Earth's surface, it might be surprising to hear that Earth kept only a "little" of the water that was delivered during accretion. The fact is that if the Earth were the size of a basketball, the ocean would be about the thickness of a sheet of paper.

Why do we think the Earth's water came from planetesimals rather than icy comets? There is a special kind of hydrogen called deuterium which has one proton and one neutron in its nucleus. Water in the ocean has a ratio of 6410 hydrogen atoms to every deuterium. This is not a very good match for comet water, where the ratio is closer to 1886 hydrogen atoms per deuterium, but it does match the ratio that has been measured in a few planetesimals. Still, when you look out at the ocean you can imagine that around a tenth of the water was delivered by comets, while the rest escaped into the atmosphere as water vapor from the hot rocky masses that composed the Earth's crust.

Section 2

FROM NOT ALIVE
TO ALMOST ALIVE

If you are reading this book hoping to learn how life began, I'm sorry to say that no one knows the answer yet. We do know something about what the Earth was like four billion years ago before life began, which was described in the first section. We also know a lot about life today and the kinds of biochemical compounds and energy sources that life requires. This knowledge lets us make some educated guesses about what organic compounds and energy sources might have been necessary for life to begin. We can check our guesswork against the organic compounds that are still being delivered to our planet in the form of carbonaceous meteorites. We can also extrapolate back in time to see what kinds of energy were likely to be available for the earliest forms of life.

I will begin with a list of the main ideas that have been put forward to explain the origin of life. The list is in approximate chronological order to provide a perspective on the scope of the research in this area right up to the present time. This is a very young field with just a few hundred scientists worldwide doing most of the work, so it is not surprising that there are conflicting ideas about how life can begin. Although I will mention names of the scientists associated with each idea, references will not be provided because they are often in scientific literature and written in highly technical language. All of the names and ideas are well known, so readers who are interested in further details can find them in internet sources. After presenting the main proposals that are circulating in origins of life research, I will then focus on a novel approach that is emerging from our studies.

Different proposals for how life began on the Earth

Panspermia

The concept of panspermia goes back 2500 years when the Greek philosopher Anaxagoras first wondered how life began. The term *panspermia* is from the Greek, meaning that the seeds of life are everywhere in the universe. In 1903, the Swedish chemist Svante Arrhenius brought panspermia to scientific attention; then cosmologists Fred Hoyle and Chandra Wickramasinghe boldly claimed that molecular clouds seen by astronomers in fact contained microbial life that would be distributed to habitable planets around new stars. The problem with these speculations is that they bypass the question of how life can begin and they don't have testable predictions.

A more useful proposal is that life began in our solar system on Mars and then was delivered to the Earth on meteorites produced when an asteroid impacted the Martian surface. We know this can happen because Martian meteorites have been identified in collections from Antarctica. If evidence is discovered that life either exists or once existed on Mars, there are numerous ways to confirm whether it was the source of life on Earth or had an independent origin. For instance, if Martian life used the same nucleic acids and proteins, the same genetic code and the same chirality as terrestrial life, it would support the possibility that life was delivered from Mars to Earth. If it had a different genetic code or used amino acids different from those in proteins, we could conclude that life on Mars had an independent origin.

Panspermia remains just an idea. Even though it might account for how life began on the Earth, it has no explanatory power in regard to the central question of how life can begin anywhere. I propose a new word that does have a testable prediction: panorganica. Most would agree that chemicals and their reactions are everywhere and some compounds would be organic chemicals if they contained carbon. A few would even have chemical and physical properties allowing them to assemble into complex structures that exhibit biological functions. The prediction is that life can arise on any habitable planet resembling the early Earth, such as Mars three billion years ago. That prediction is now being tested by the robotic rovers looking for signs of life on our sister planet.

Coacervates

Alexander Oparin wrote the first scientific book about the origin of life in 1924, in which he proposed that life's origin could be understood as a chemical process. His book was written in the Russian language and was not widely distributed, but in 1938 an English translation became available. In 1929 the British scientist J. B. S. Haldane published a brief essay describing an idea similar to Oparin's: that life could emerge as a result of chemical reactions occurring on the early Earth. Oparin continued to perform research over the next fifty years. During that time he came up with the concept of coacervates, defined as self-assembled, cell-sized structures composed of polymers. This concept inspired Sidney Fox to heat dry amino acids to temperatures associated with volcanos. The amino acids melted and formed polymers he called proteinoids. Furthermore, under certain conditions the polymers aggregated into proteinoid microspheres that he believed were steps toward the origin of life. As we learned more about the details of molecular biology, this approach has been largely abandoned.

Electric sparks and gas chemistry

In 1952, Stanley Miller was a graduate student at the University of Chicago, working under the supervision of Nobelist

Harold Urey. He convinced Urey that it might be interesting to see what happened if a mixture of gases was exposed to an electric spark. The idea was that electrical discharges such as lightning might drive chemical reactions in the early atmosphere before life began. They assumed that the atmosphere would be composed of hydrogen, methane, ammonia, and water vapor, a composition meant to simulate the primordial atmosphere of the early Earth and contained in a glass sphere. After several days of sparking, the mixture became reddish-brown, making it obvious that something was happening. When Miller analyzed the water solution, the amazing result was that several amino acids had been synthesized! The publication of Miller's paper in 1953 caused a sensation and is generally considered to mark the beginning of the scientific study of the origin of life.

Miller spent most of his career in the Department of Chemistry at the University of California, San Diego. He and his students published hundreds of papers, mostly confined to reactions that produced small molecules related to the origin of life rather than polymerization reactions and assembly of boundary membranes.

Mineral surfaces

In his 1967 book *The Origin of Life*, John Dexter Bernal suggested that clay minerals have special properties that might guide chemical reactions related to the origins of life. This idea was explored further by the Scottish researcher Graham Cairns-Smith, who proposed that the surfaces of clay minerals actually contained information in their crystalline organization that could be passed on to nucleic acids that assembled on the surface. James Ferris spent much of his career at the Rensselaer Polytechnic Institute investigating laboratory models of clay chemistry, and he showed that a special clay called montmorillonite adsorbs chemically activated mononucleotides in such a way that they polymerize into short strands of RNA.

Günter Wächtershäuser proposed that the surface of an iron sulfide mineral called pyrite (commonly known as fool's gold) has the potential to adsorb organic compounds related to life. Furthermore, the formation of pyrite mineral was associated with the ability to provide reducing power that could potentially transform simple molecules like carbon dioxide into biologically useful compounds like amino acids. In other words, the first forms of life did not begin as cells but instead as metabolic reactions on the surface of a common mineral. The two-dimensional metabolism was later encapsulated by lipid membranes to become the first forms of cellular life. Wächtershäuser undertook some experimental tests of his idea and showed that a few simple chemical species could be synthesized, but the concept has not advanced much further.

Hydrothermal vents

The hydrothermal vents called black smokers were discovered by Jack Corliss in 1977 while diving in the Alvin submersible near the Galapagos Islands (Plate 9). The heat source for black smokers is an underlying magma field, so the acidic vent water can be very hot, exceeding 300°C. The black "smoke" is composed of metal sulfides. Another kind of hydrothermal vent called Lost City was discovered in 2001 near the mid-Atlantic ridge, between the continental land masses of Africa and the Americas. These are referred to as white smokers, and instead of hot magma the heat source is serpentinization caused by sea water reacting with sea floor minerals to form another mineral called serpentine. The reaction produces highly alkaline water with dissolved hydrogen gas.

Both types of vents support microscopic bacteria and larger organisms such as tube worms. Soon after they were discovered, it was proposed that life may have begun in hydrothermal vents because of the chemical energy available in aqueous solutions. This idea was expanded into a conceptual network of reactions that will be described later. However, because the vents exist in deep marine conditions, no experiments have been undertaken to test the ideas directly.

Metabolism first model

Metabolism is defined as all the biochemical reactions that occur in a living cell which are essential for life. It is obvious that a primitive version of metabolism must have become incorporated into the first forms of life, but it is not obvious how this could have occurred in the absence of enzymes and transport of nutrients across membranes into cells. However, we can make some guesses by extrapolating from what we know about metabolism in living cells today, and the origin of metabolism continues to be a focus of attention for some of the scientists working on the question of how life can begin. For instance, Michael Russell, William Martin, and Nick Lane have proposed that hydrothermal vents may have provided a source of chemical energy for primitive metabolic reactions, and Harold Morowitz argued that remnants of primitive metabolic pathways can still be observed in life today.

Lipid World and the GARD model

Doron Lancet at the Weizmann Institute in Israel suggested that an important step toward life would be the self-assembly of certain compounds into microscopic structures having specific sets of organic compounds. If the mixture of compounds could extract energy from the environment, they might change in such a way that the structures grow and reproduce by splitting into daughter cells. The main point is that their molecular composition represented a kind of information to be passed along to progeny. The concept was first modeled computationally in a program called the GARD model (graded autocatalytic replication domain). The assembly of lipid molecules into microscopic aggregates called micelles or vesicles with a membranous boundary structure is considered to be a real-world model of GARD. The concept was expanded into a more general concept of a Lipid World that preceded the origin of life and used compositional information rather than sequences of monomers in linear nucleic acid molecules that store genetic information in all life today.

RNA World

As knowledge of nucleic acid structure slowly accumulated after the structure of DNA was reported by James Watson and Frances Crick in 1953, it was speculated that RNA might have catalytic properties. This was confirmed in 1982 when Tom Cech and Sid Altman discovered that certain kinds of RNA could catalyze their own hydrolysis, for which they won the Nobel Prize in 1989. Catalytic RNA species became known as ribozymes because they resemble enzymes but are composed of RNA instead of proteins. In 1986, Walter Gilbert at Harvard wrote a brief essay in which the term *RNA World* was used to define early forms of life that use RNA both as a catalyst and to store genetic information. This has been a fruitful working hypothesis and will be described in further detail later.

Warm little ponds

In a letter written in 1871 to his friend Joseph Hooker, Charles Darwin included a few prescient sentences that have become current again nearly 150 years later.

It is often said that all the conditions for the first production of a living being are now present, which could ever have been present. But if (and oh what a big if) we could conceive in some warm little pond with all sort of ammonia and phosphoric salts,—light, heat, electricity &c present, that a protein compound was chemically formed, ready to undergo still more complex changes, at the present such matter would be instantly devoured, or absorbed, which would not have been the case before living creatures were formed.

Does a testable hypothesis emerge from Darwin's conjecture? Other researchers in the field will agree with much of the information to follow in this section, but some will be skeptical about ideas that don't match their own. In a nutshell, the freshwater in hot springs on volcanic land masses may have physical and chemical properties that are more conducive to life's origin than salty seawater. The reasoning comes from an observation anyone can make if they visit volcanic landscapes, which is that the small pools fed by hot springs go through cycles of evaporation and refilling. Laboratory experiments have shown that wet-dry cycles concentrate potential reactants and provide energy for synthesizing the essential polymers of life. If soap-like molecules are present, the polymers are encapsulated in microscopic compartments bounded by membranes. This process was described in detail in my book *Assembling Life* (Oxford University Press, 2019) and has been tested in the laboratory and in volcanic sites like Yellowstone, Kamchatka, and New Zealand hot springs.

The encapsulated polymers are called protocells. They are not alive, but they do have the capacity to undergo selection and the first steps of evolution toward life. This is speculative and does not yet have experimental support, but the testing process has begun. When the results are available, they might show that a system of functional polymers in protocells can take up energy and nutrients from the environment, using them to grow and reproduce. Even with positive results we could not conclude that life did begin this way four billion years ago, but we would be able to state that this is how life can begin on habitable planets like the early Earth and Mars.

The rest of this section describes results that are consistent with a hot spring origin, as well as the main ideas of the Lipid World and RNA World.

All life is cellular, and probably the first forms of life as well

The unit of all life today is the cell, but how did the first life become cellular? Is non-cellular life possible? Imagine that you are a young chemist who has just been hired as an assistant professor at a small college. The chair of the department takes you to view your new laboratory. There is plenty of space and some nice shelves and lab benches, but little piles of various chemicals are arrayed on the benches. The chair is embarrassed to tell you that the department ran out of funds and could not afford glassware. "But how can I do experiments?" you ask.

In fact, in the absence of compartments like flasks and beakers and test tubes, experiments are virtually impossible. The same is true for the origin of life. Unless mixtures of various soluble compounds can be held together in one place, the natural experiments required for life to begin could never occur.

So, where did the first membranes come from? This is actually one of the easier questions to answer because we know a lot about the membranes that assemble in living cells today. If we extract the membranes of any cell, from the simplest bacteria to the membranes of neurons in the brain, we discover that they are composed of phospholipids mixed with other molecules such as cholesterol. If we mix the lipids with water, they spontaneously form microscopic cell-sized vesicles. If we treat the phospholipids with acid, they fall apart into soap-like fatty acids, glycerol, and phosphate. Fatty acids are in fact the same molecules that produce the membranes of soap bubbles.

How do we know?

A soap molecule has a long hydrocarbon chain that ends in a carboxyl group similar to carbon dioxide (CO_2). A typical soap looks like this:

$$H_3C-CH_2-CH_2-CH_2-CH_2-CH_2-CH_2-CH_2-CH_2-CH_2-$$
$$CH_2-COOH.$$

This one is lauric acid and falls into a class of compounds called amphiphiles, from Greek words meaning "love both." The hydrocarbon chain "loves" oil and the carboxyl group "loves" water, which gives such molecules an important property that is essential for life. Everyone has blown bubbles composed of amphiphilic soap: the colorful boundaries of the bubbles are in fact a kind of membrane. Other amphiphiles compose the membranes that are essential boundary structures for all cellular life today, as well as the first primitive cells that emerged by self-assembly on the early Earth.

Figure 2.1 shows what happens when a tiny bit of soap is dried on a glass slide and covered with a cover slip. Nothing happens until water is added, and then the soap molecules immediately begin to form tubular structures from which cell-sized vesicles begin to bud off.

The amazing thing is that the same process can occur in organic material extracted from a carbonaceous meteorite, as shown in Figure 2.2. The extract was dried on a microscope slide followed by addition of a drop of water. Abundant microscopic vesicles assembled from the soap-like compounds known to be present in the extract. They light up when exposed to ultraviolet light because fluorescent compounds called polycyclic aromatic hydrocarbons (PAH) are also in the membranes. From results like this, it is reasonable to assume that similar mixtures of organic compounds accumulated on volcanic land masses of the prebiotic Earth. They would have been flushed by rainfall into hot springs in which the soap-like molecules formed the microscopic compartments required for the origin of cellular life.

Life requires liquid water

To get a perspective on the necessity for liquid water, we can ask the opposite question: Why can't life occur in ice? Or in a place like the Atacama Desert in Chile which receives almost

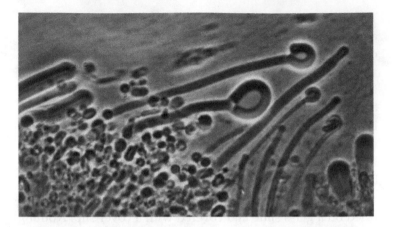

Figure 2.1 The thick membranes that assemble when dry soap molecules are exposed to water are actually composed of hundreds of layers that form the tubes and vesicles in the image. These are unstable and slowly break up into microscopic membranous vesicles that have just a single bimolecular layer of soap molecules, too small to be visible at the magnification used in the photo.

Credit: Daniel Milshteyn

no rainfall? It is true that living organisms can survive in a frozen or dry state, but are they alive? Not exactly, because they exhibit none of the usual functions that define life such as metabolism, growth, and reproduction.

This is why hydrothermal pools are a plausible site for life's origin. There are several reasons, but an important feature of the pools is that they undergo cycles of hydration and dehydration. Some cycles can be very rapid, such as water erupting from geysers and then splashing and drying on nearby hot rocks. Other cycles occur more slowly as pools evaporate over time intervals of days to weeks, and then are refilled by precipitation. Two kinds of reactions can occur in the films that are deposited on mineral surfaces during evaporation: self-assembly of amphiphiles into membranous structures and condensation reactions that produce polymers.

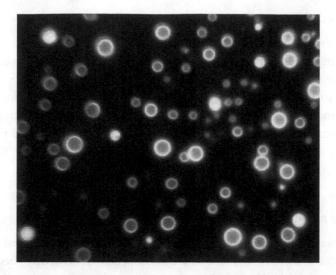

Figure 2.2 Amphiphilic compounds isolated from the Murchison meteorite can assemble into cell-sized vesicles bounded by a membrane. The compounds are a mixture of fatty acids resembling soap mixed together with polycyclic compounds that produce a blue fluorescence when illuminated with ultraviolet light under the microscope. Even though they are likely to be older than the Earth, the compounds retain the ability to assemble into membranes. The smaller vesicles are about the size of red cells in blood.

Credit: Author

Now we can answer the question of why life needs liquid water. Liquid water is a benign solvent within which diffusion of molecular solutes is possible. Diffusion in this case has a technical meaning, which refers to the motion of dissolved molecules that move randomly in the mixture. If a concentration gradient is present, diffusion causes solutes to move from regions of greater concentration to regions of lower concentration. For example, when we breathe in, there is a higher concentration of oxygen in the air than in the blood circulating through the lungs, so oxygen diffuses into the blood, which transports it to the rest of our body. If we plant a seed, water diffuses out of the soil into the dry seed which then can spring to life and begin to grow.

How do we know?

It seems obvious that life needs liquid water because we see living organisms wherever there is water in the form of rivers, lakes, and oceans, or rainfall on land. But that's just an observation and doesn't answer the "why" question. Think for a moment: Is it possible to blow soap bubbles with dry soap? Of course not! Water is essential to dissolve the soap, and the physical properties of water and soap molecules allow the soap to spontaneously form membranous vesicles.

Life probably began in freshwater on volcanic islands

I need to warn readers that this heading is controversial. Because there is so much water in the ocean, it has always been assumed that life must have begun there. However, when this assumption is examined more closely, two significant concerns arise. For instance, the volume of the ocean is huge, so solutions of the organic compounds required for life to begin would be so dilute that they could not find each other to react. In contrast, even dilute solutions become extremely concentrated when water evaporates from small pools of freshwater in volcanic hot springs. Another concern is that seawater is what we call hard water because of the calcium and magnesium ions that are present. If you try to wash your hands in hard water with soap, it doesn't work very well. The reason is that the calcium and magnesium react with the soap and cause it to aggregate into clumps rather than remain "soapy." However, we have found that membranous vesicles readily assemble in the freshwater of volcanic hot springs.

Finally, and probably most important, it's impossible for monomers to form polymers in seawater because the reaction is thermodynamically uphill. In other words, energy is required for polymers to form, and the only way to put energy into polymerization reactions in solution is to chemically activate the monomers. Metabolic processes in all life today activate monomers like amino acids and use enzymes to polymerize them into proteins, but no one has experimentally demonstrated a plausible mechanism by which this could occur in the prebiotic ocean. In contrast, it has been known for years that sufficient energy can be introduced in freshwater simply by evaporating a solution of monomers and heating the dry film.

How do we know?

In our research we have placed small vials containing the monomers of RNA into the edge of a New Zealand hot spring and put them through four wet-dry cycles using water from the spring. When the results were analyzed in the laboratory, very good yields of polymers resembling RNA were found. This result supports the idea that the nucleic acid polymers of life can be synthesized in conditions resembling those of the prebiotic Earth and not just in the laboratory. Plate 10 shows an artist's rendition of what the Earth was like at the time that life began.

Life needs monomers

Three main kinds of molecules called monomers can be chemically linked into linear or branched strands called polymers. The monomers are amino acids, carbohydrates, and nucleotides; examples are shown in Plate 11. It is important to understand that the chemical bonds that link monomers into polymers are formed by a reaction called condensation in which a water molecule is removed from between chemical groups of the monomers. This sounds complicated, but it is actually simple. For instance, one of the most important linkages in biology is called an ester bond. Suppose you mixed acetic acid, which is what makes vinegar sour, with ordinary alcohol. Some of the acetic acid will form an ester bond with the alcohol to produce ethyl acetate and a water molecule:

$CH_3–CH_2–OH + CH_3–COOH \rightarrow CH_3–COO–CH_2–CH_3 + H_2O$. Many of the flavors and aromas in fruit and vegetables are esters, such as benzyl acetate in pears and strawberries, and butyl butyrate in pineapples. Methyl salicylate is the flavor of wintergreen, and amyl acetate gives bananas their aroma.

The point is that if monomers continue to be added until they form a chain, a polymer has been synthesized. One well-known polymer is the polyester used in clothing, which is composed of organic acids and alcohols held together by ester bonds. Nucleic acids like DNA and RNA also fit the definition of a polyester.

How do we know?

When scientists began to study the chemicals of life in the 1800s, they found that most of the substances they isolated could be broken down by heating them in solution with an acid such as hydrochloric acid. For instance, heating starch with an acid caused it to break down into the molecules now called glucose. A similar treatment of proteins caused them to fall apart into amino acids. The first nucleic acid was described in 1869, and again acid treatment caused it to break up into four different monomers called nucleotides.

Life today uses twenty different amino acids to make proteins, but what was the source of amino acids used by the first forms of life? We now know that the Murchison meteorite contains over seventy compounds classified as amino acids, which means that amino acids can be synthesized by non-biological chemical reactions. The Murchison meteorite also contains nucleobases such as adenine and guanine that are part of nucleic acid structure. Nucleotides are more complicated than amino acids because they are composed of a nucleobase linked to a sugar which is linked to a phosphate. However, a series of chemical reactions was recently reported that can synthesize all four nucleotides from simpler compounds likely to be available on the prebiotic Earth. Could nucleotide monomers have formed the polymers of nucleic acids on the early Earth? It seems possible that life did not invent nucleic acids, but instead emerged when nucleic acids were encapsulated in membranous compartments during wet-dry cycles occurring in hot springs.

Life is composed of polymers

Polymers are familiar in today's world because they compose plastics such as polyethylene, polystyrene, polypropylene, polyesters, and the list goes on. Life is not composed of plastic, but a living cell instead uses polymers called nucleic acids and proteins. Figure 2.3 illustrates what a protein polymer looks like at the molecular level.

A simple way to illustrate how proteins and nucleic acids function is to imagine constructing a living cell. No one has succeeded in doing this yet, but Figure 2.4 shows how a bacterial cell could be assembled, in a thought experiment. We will start with DNA in the upper left, which contains the genes that guide protein synthesis. A typical bacterial DNA is a ring wound up into a structure called a nucleoid, which contains around 5000 genes embedded in a nucleic acid polymer composed of five million base pairs. For comparison, the DNA of

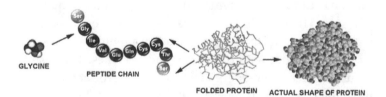

Figure 2.3 Amino acids like glycine can be incorporated into long polymeric strands of proteins. The strand then folds into a specific structure that has physical and chemical properties related to cell functions. One of the most important is enzymatic catalysis of metabolic reactions which take place on the surface of the enzyme where specific amino acids form an active site.
Credit: Adaptation by author.

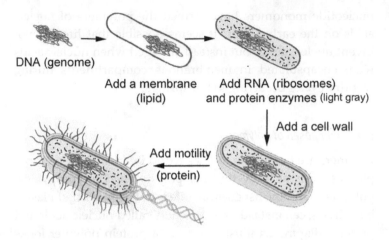

DNA (genome)

Add a membrane
(lipid)

Add RNA (ribosomes)
and protein enzymes (light gray)

Add a cell wall

Add motility
(protein)

Figure 2.4 A step-by-step recipe for assembling a living cell.
Credit: Author

the human genome has three billion base pairs. The next step is to put the DNA into a membranous compartment. This is easy to do in the laboratory and the resulting structures are called protocells. These are not alive, but they are an essential step toward life. Then we add RNA and proteins. Most of a cell's RNA is in ribosomes, the black dots in the illustration. Proteins are also part of the ribosome structure, and thousands of protein enzymes in the cytoplasm are shown in light gray. The resulting structure is now alive because it is a system of genes in DNA encapsulated in a membranous compartment together with ribosomes and enzymes. The system can use nutrients and energy to grow and reproduce. Some of the simplest microbial cells are just like this, but they are very fragile and only grow in special conditions. Most free-living bacteria have evolved a cell wall that protects them from environmental stresses. Some bacteria even have flagellae composed of proteins. These spin like little propellers and move bacteria through water so that they have a better chance of finding nutrients.

Plate 1 Astronomers now have enough information to make a map showing how galaxies are distributed in the visible universe. They are not random, but instead gather into clusters and strands. The white band is our own galaxy, the Milky Way, observed edge on.

Credit: WISE, 2MASS

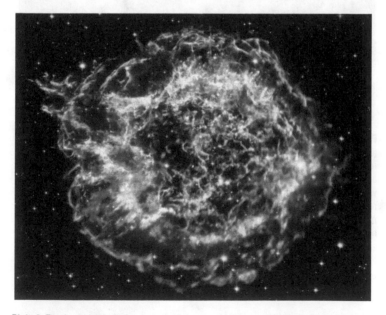

Plate 2 The tiny white dot in the center is all that remains of a star that reached the end of its life and exploded as a supernova called Cassiopeia A. Elements like iron (purple), sulfur (yellow), calcium (green), and silicon (red) compose the dust particles ejected by the explosion. They emit X-rays as well as visible light, and color has been added to indicate the distribution of elements in this combined image from the Chandra X-ray Observatory and the Hubble telescope.

Credit: NASA, CXC. SAO, NASA STSxl

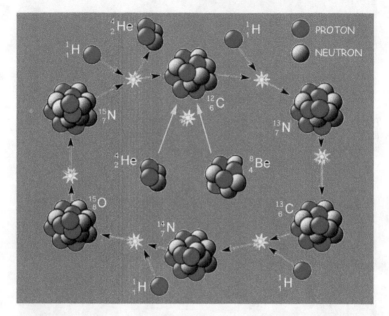

Plate 3 This is a cycle called stellar nucleosynthesis that produces carbon (C), oxygen (O), and nitrogen (N), the primary elements of life along with hydrogen (H). The red spheres represent protons and the gray spheres neutrons. The lower numbers to the left of each abbreviation indicate the atomic number defined by the protons in the nucleus; the atomic weight that includes the neutrons is the upper number. Each yellow burst represents a fusion reaction that releases energy, most of which results from the fusion of hydrogen atoms to produce helium. When a helium nucleus with two protons and two neutrons fuses with a beryllium nucleus having four protons and four neutrons, the result is a carbon nucleus with six protons and six neutrons. The carbon can then undergo further fusion events with protons to produce nitrogen and oxygen.

Source: Adapted by author from public domain sources.

Plate 4 The galaxy called NGC 1566 (Spanish Dancer Spiral Galaxy) has intrinsic beauty. The jewel-like pink regions are where new stars are forming, and the dark areas are clouds of interstellar dust ejected into space when old stars collapse and explode.

Credit: NASA, ESA, Hubble; processed by Leo Schatz

Plate 5 New stars emerge from clouds of interstellar dust and gas, the ashes of earlier stars that exhausted their supply of fusion energy and exploded.

Credit: NASA, JPL-CalTech, WISE

Plate 6 A molecular cloud in Aries. The inset shows an actual interplanetary dust particle captured by a high-flying aircraft. Such particles are the remains of the interstellar molecular cloud in which our solar system formed. The light blue tint was added to indicate the thin layer of ice coating its surface. It has been estimated that 30,000 tons of interplanetary dust particles (IDP) accumulate in the upper atmosphere every year and float downward until they reach the surface. Even four billion years after the primary accretion was completed, the Earth is still gathering extraterrestrial matter in the form of dust particles and meteorites.

Source: Adapted from published Hubble telescope images

Plate 7 A new telescope in Chile called the Atacama Large Millimeter Array can actually see what appears to be a developing solar system in a nearby star called HLTauri. The star is only a million years old and is surrounded by a disk of gas and dust, just as predicted by theory. The obvious gaps in the disk are most likely being produced by new planets as they gather up the dust. It is reasonable to assume that our solar system was formed by a similar process.

Credit: ESO, ALMA

Plate 8 The Earth-moon system resulted from a collision 4.4 billion years ago between the early Earth and another planet when their orbits happened to intersect. The moon was produced from the ejected debris that formed a transient ring around the Earth. At first, the moon was much closer than it is today, as shown in this artist's image. So much energy was released by the collision that the newly formed Earth and moon were molten.

Credit: Mark Garlick

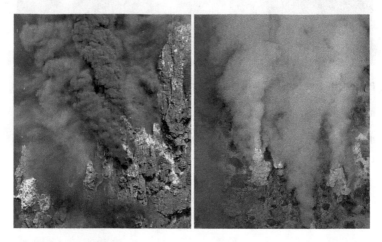

Plate 9 Two kinds of hydrothermal vents are shown, black smokers on the left and "white smokers" or alkaline vents on the right.

Credit: USGS

Plate 10 There were no continents when life began four billion years ago. Instead, volcanic land masses similar to Hawaii and Iceland had emerged from a salty global ocean. Water evaporated from the ocean and rained onto the volcanic islands, then formed freshwater hot springs and pools resembling those found in Yellowstone today.

Credit: Ryan Norkus and Bruce Damer

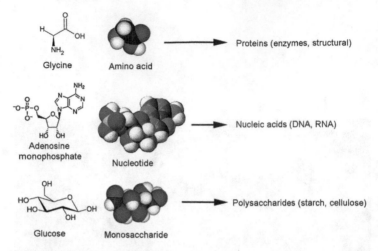

Plate 11 Chemical structures of three primary monomers of life: an amino acid (glycine); a nucleotide (adenylic acid); and a carbohydrate (glucose).

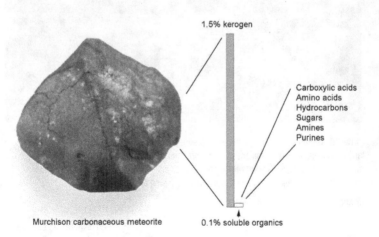

Plate 12 Organic compounds are present in carbonaceous meteorites such as the one that fell near Murchison, Australia, in 1969. Carbonaceous meteorites contain 1–2 percent of their mass as an insoluble organic substance called kerogen and smaller amounts of soluble compounds that have chemical properties related to the origin of life.

Credit: Author

Plate 13 An evaporating hot spring pond near Mount Mutnovsky in Kamchatka, Russia.
Credit: Author

Plate 14 Fluorescently stained DNA being encapsulated in lipid vesicles.
Credit: Author

Plate 15 A stromatolite fossil in the Pilbara region of Western Australia. The mineral layers were built up by microbial films that accumulated on their surfaces over three billion years ago.

Credit: Bruce Damer

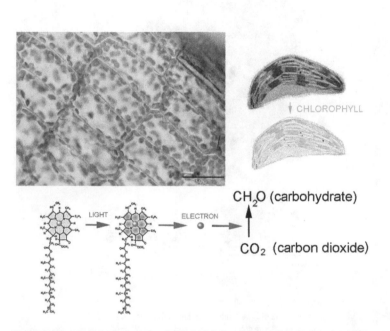

CH$_2$O (carbohydrate)

CO$_2$ (carbon dioxide)

LIGHT   ELECTRON   CHLOROPHYLL

Plate 16 Chloroplasts in plants capture light, which is the primary source of energy for all life on Earth. When light is absorbed by a green chlorophyll molecule, its excited state releases an electron, which is used to reduce carbon dioxide to carbohydrates. Another molecular system takes electrons from water to replace the electrons lost from chlorophyll when light is absorbed. The oxygen that is produced is a source of energy for all aerobic organisms.

Credit: Author

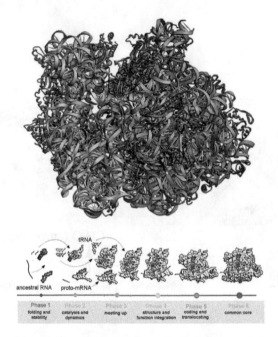

Plate 17 Ribosome structure: The small subunit is composed of one molecule of RNA (blue) interacting with twenty-one proteins (purple); the large subunit is composed of two RNA molecules (gray) and thirty-one small proteins (also in purple). Two other RNA molecules involved in protein synthesis include mRNA, which carries genetic information from DNA to the ribosome, and tRNA, which carries amino acids to the active site where they are attached to a growing protein chain. The possible stages of ribosome evolution have been deduced by determining bases sequences in ribosomal RNA. Some are highly conserved and are therefore considered to be ancient, while other sequences vary significantly and are likely to be more recent additions.

Credit: Harry Noller and Loren Williams

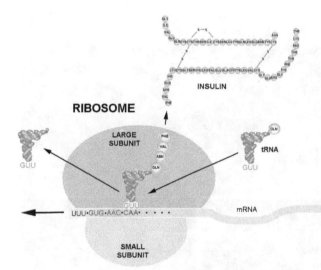

Plate 18 This figure shows a ribosome synthesizing a small protein called insulin which is composed of two strands held together by disulfide linkages. Messenger RNA (mRNA) carries a sequence of bases copied from DNA in the nucleus of a pancreatic cell and is moving from right to left through the ribosome. There are four bases in mRNA—adenine (A), uracil (U), guanine(G) and cytosine (C)—which are present as groups of three called codons. Each codon is specific for one of the twenty amino acids that are the monomers of proteins. Another form of RNA called transfer RNA (tRNA) carries amino acids to the ribosome. In the illustration, a tRNA molecule is shown with the base sequence GUU at one end and the amino acid glutamine (GLN) on the other end. When the tRNA reaches a ribosome, the GUU binds to the matching triplet CAA on the mRNA and the amino acid is added to a growing peptide chain. Four amino acids have been added so far—Phe (phenylalanine) Val (valine), Asn (asparagine) and Gln (glutamine)—and the same amino acids at the end of the longer chain in the completed insulin molecule. *Author*

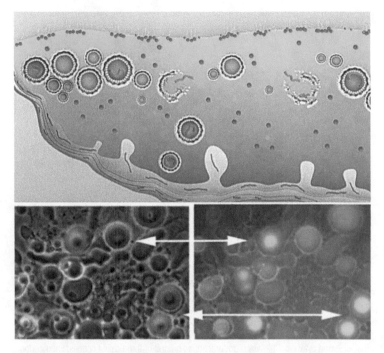

Plate 19 Membranous compartments are shown budding off lipid layers on a mineral surface. Polymers (red) have been synthesized within the layers during the dry phase of a wet-dry cycle, and these become encapsulated in lipid vesicles to form protocells. In the aqueous phase, some protocells are disrupted but others survive because they are stabilized by the encapsulated polymer. The two lower figures show lipid vesicles that were produced by this process in which just one nucleotide was cycled in the presence of the lipid to synthesize DNA. After four cycles the protocells contain encapsulated DNA stained with a fluorescent dye.

Credit: Art, Ryan Norkus; photo, author

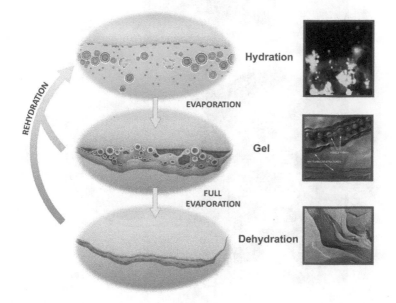

Plate 20 A unified view of the dry-wet-moist cycle made possible in hot spring ponds, in which three distinct phases subject populations of protocells to combinatorial selection: a dry phase promotes synthesis of polymers; a wet phase buds off collections of these polymers into protocells and tests them for stability and longevity; in the intermediate moist gel phase, aggregates of protocells called progenotes assemble. Progenotes can share polymers that add survival value to the colony. They represent a unit of selection which supports primitive metabolism, growth by polymerization, catalyzed replication of the polymers, and finally the transition to living cells. Images on the right show microscopic evidence supporting each of the three stages.

Credit: Art, Ryan Norkus; microscopic images, author

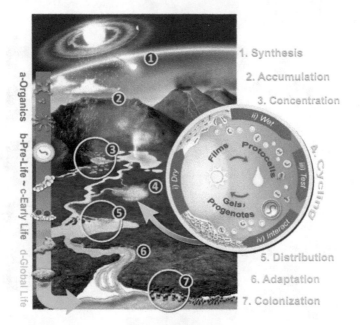

Plate 21 The Big Picture begins with delivery of organic compounds to a volcanic land mass in Phase 1, where they accumulate on mineral surfaces. Phases 2 and 3: Precipitation flushes the compounds into hot spring pools, where they can be concentrated during wet-dry cycling. Phase 4. If monomers are present together with amphiphilic compounds, the monomers become polymerized within the matrix of lipid layers; then in the wet phase lipid vesicles emerge as protocells with encapsulated polymers. Phase 5. Protocell aggregates are distributed as primitive progenotes that undergo selection for increasing robustness, catalyzed metabolism, and functions such as photosynthesis. Phase 6. The progenotes are distributed downhill toward the sea and adapt to increasingly salty water. Phase 7. By this time the progenotes have all the functions required for life and represent the last universal common ancestor, LUCA.

Credit: Ryan Norkus

How do we know?

The electron microscope was invented in the 1930s and gave us our first view of cell structure at the molecular level of resolution. The middle image in Figure 2.5 shows a single bacterium that was stained with a heavy metal called osmium, and then embedded in epoxy plastic and sliced into very thin sections with a diamond knife. The outer cell wall and membrane are clearly visible and also the nucleoid DNA in the center. The image on the left was taken with a scanning electron microscope that reveals the bacterial cell surfaces as three-dimensional structures. The right-hand image is a stained bacterial cell that was dried on a very thin membrane and shows structures called fimbria that protrude from the cytoplasm through the cell wall. Images like these let us draw the diagram shown in Figure 2.5. The important point of this

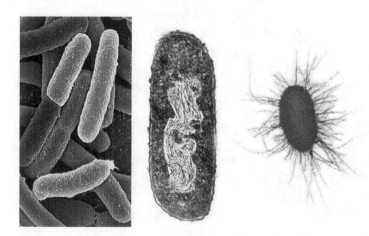

Figure 2.5 The image on the left shows what several bacteria look like in three dimensions when viewed by a method called scanning electron microscopy. The center image shows a single bacterium that was fixed and stained, and then embedded in plastic and sliced into a thin section to reveal the interior structure, such as the DNA in the nucleoid. The image on the right is a bacterial cell stained by a technique that reveals fiber structures that allow bacteria to adhere in colonies and on surfaces.

Credit: Prepared by author from public domain images

exercise is that with the exception of membranes, every structure added to the diagram is a polymer, and the polymers are constructed from monomers like amino acids, nucleotides, and carbohydrates. This means that to understand the origin of life, we need to find a way for polymers to have been synthesized non-enzymatically four billion years ago, and then encapsulated in membranous compartments to form protocells capable of evolving into the first living cells.

Organic compounds were available to support the origin of life

The Earth, its water, and its atmosphere are composed of compounds delivered by accretion of material present in the original solar nebula that surrounded the sun. However, shortly after the Earth had nearly reached its original size it collided with a Mars-sized planet that happened to cross its orbit. The collision produced a ring of rocky material around the Earth from which the moon formed. The energy released by the collision left the moon and Earth at the temperature of molten lava. Only simple carbon compounds like carbon dioxide could survive this temperature, so more complex compounds would be destroyed. The organic compounds required for life to begin could only exist after the Earth cooled and a global ocean was present.

There are two possible sources of organic compounds: delivery by microscopic dust, meteorites, and even comets that was continuing at a very high rate four billion years ago; or synthesis by chemical processes occurring in the Earth's atmosphere or crust. It is uncertain which was the primary source, but in either case the compounds would have immediately begun to be chemically transformed by energy sources such as heat and light. This means that the mixture would not be stable, but instead there would be a constant input of fresh organic compounds that were being continuously transformed into other compounds. We do know that organic compounds are still being delivered today when carbonaceous meteorites and dust particles fall to Earth, so we can use that as a guide to the kinds of organic compounds likely to be available for the chemical reactions leading to the first forms of life. We also know that important organic compounds like amino acids, nucleobases, and lipid-like hydrocarbons can be produced in the laboratory by simulating prebiotic conditions. A reasonable assumption is that such reactions were also occurring on the early Earth.

How do we know?

In September 1969, a fireball blazed across the sky above Murchison, Australia, and exploded. Pieces fell into the neighboring fields and ~100 kg were collected by townspeople and scientists who rushed to the scene. One fragment was delivered to NASA Ames in Mountain View, California, where it was analyzed by modern techniques. The Murchison meteorite contained numerous organic compounds including amino acids, and the compounds must have been synthesized by non-biological chemical reactions. The next fifty years of research added other compounds to the list of those that were likely to be available on the early Earth before life began, including the nucleobases of DNA and RNA, carbohydrates like ribose that are also part of nucleic acids, and long-chain hydrocarbon derivatives called fatty acids that can assemble into membranes. The truly astonishing thing is that all of these compounds can be synthesized from simple reactive compounds like hydrogen cyanide (HCN), formaldehyde (HCHO), and carbon monoxide (CO).

Plate 12 shows the composition of the organic compounds in the Murchison meteorite. The next question was to find out how they can assemble into the essential biopolymers of life—proteins and nucleic acids—and how the biopolymers can be encapsulated in membranous compartments as an essential step toward the origin of life.

In order to react, organic compounds must be concentrated

Organic chemists know that a solution of potential reactants must be concentrated to make a reaction possible. Any compounds falling into the global ocean on the early Earth would become extremely dilute. Even if all of the amino acids and carbohydrates in life today were dissolved in the ocean, the concentration would be so dilute that each molecule would be surrounded by ten million water molecules. However, there is an alternative to seawater, which is the freshwater evaporating from the salty ocean and falling as precipitation on volcanic land masses. Anyone who visits the volcanic island of Hawaii experiences this almost every day when it rains. The important property of freshwater on land is that it forms small pools, which undergo periodic cycles of drying. Any organic compounds that get flushed into the pools become extremely concentrated during the drying process and form films on the mineral surfaces. Chemical reactions can occur in the concentrated films, including polymerization that increases the complexity of otherwise simple solutions of organic compounds.

How do we know?

Anyone who visits volcanic sites like Yellowstone National Park in Wyoming or Rotorua in New Zealand can see evidence for wet-dry cycles. The image in Plate 13 shows evaporating puddles fed by rainfall and hot springs on the flank of Mount Mutnovsky in Kamchatka, Russia. Dried material forms "bathtub rings" on all of the rocks and then dissolves again when it rains. Experimental studies have shown that polymerization reactions can occur in such rings.

Energy and life's beginning

The concept of biological energy can be understood intuitively, and so can two related words called enthalpy and entropy. Everyone knows that exercise requires us to use energy and get hot. The heat is related to enthalpy because when energy in the molecule ATP (adenosine triphosphate) causes muscle contraction and relaxation, heat is given off by the reaction. We also know that as a general rule if we put things in order they tend to become disordered as time passes. The disordering effect is related to entropy. An example of entropy is what happens when we put a salt crystal into water. The sodium and chloride atoms that compose salt are highly organized in the crystal but become disorganized when they dissolve in the water. In other words, entropy increases.

But this intuitive understanding becomes much more complex when we try to measure energy, and particularly when we try to understand how it controls the chemical reactions and physical processes related to life. We can start by analyzing a chemical reaction that everyone is familiar with: lighting a candle. We strike a match, hold it to the wick, and a reaction begins in which the hydrocarbon molecules in the candle wax react with oxygen in the air to produce a flame. Heat is produced as a result, and this can be measured by letting the flame warm some water while using a thermometer to measure the increase in temperature. If we use a liter of water, under ideal conditions every gram of wax that burns increases the temperature of the water by 9°C. The definition of a kilocalorie is the heat energy required to raise the temperature 1°C, so candle wax has 9 kcal per gram, which is the same energy content as one gram of fat. It's an interesting fact that the human body burns fat at about the same rate as a burning candle, so with each breath we lose a few milligrams of carbon dioxide that used to be present in the hydrocarbons of fat stored in our bodies.

To summarize, the candle wax contains potential energy in its hydrocarbons that can be released when it reacts with oxygen. The products of the reaction are heat energy, carbon dioxide, and water. Every cell in all life on Earth does something similar if it uses oxygen as an energy source to "burn" fat or carbohydrates and release carbon dioxide. But what about entropy? Where does that come into the picture? And how did the first forms of life get started in the absence of oxygen in the atmosphere?

Every chemical reaction that is spontaneous can be described by two measurable quantities related to a change in enthalpy and a change in entropy. The enthalpy change is measured by the heat given off by the reaction and the entropy by a change that can be loosely described as order proceeding toward disorder. A chemical reaction is spontaneous if the changes in enthalpy and entropy are both favorable, in other words, if heat is given off by the reaction and the products of the reaction are more disordered than the reactants. In the case of the candle burning, the loss of chemical energy to heat is obvious, but the change in entropy is due to the fact that all of the original hydrocarbon molecules were organized within the candle wax, and then become disordered as they are released as carbon dioxide and diluted into the air in the room.

Most of the reactions of life are dominated by the enthalpy term, but there is one important reaction that is not. Imagine that you could somehow mix soap molecules into water and watch what happens to them. At first, they are present as individual molecules surrounded by water molecules, but then over time they assemble into clusters called micelles. Each micelle contains hundreds of molecules with their hydrocarbon tails inside and hydrophilic carboxyl groups on the outside. Then if you continue to add more soap the micelles begin to coalesce into beautiful cell-sized vesicles. Even though this is a spontaneous reaction that produces order from disorder, if you measured the temperature there would be almost no change, which means the micelles and vesicles are stabilized by a favorable change in entropy rather than enthalpy.

How do we know?

Water molecules are held together as a liquid because there are weak electrical charges on the oxygen and hydrogen atoms of H_2O. The oxygen is negatively charged and the hydrogens are positively charged. The charges produce weak attractive forces called hydrogen bonds which maintain the water as a liquid at ordinary temperatures. In other words, water molecules are like little bar magnets with a north and south pole, and if you mixed lots of bar magnets they would all assemble so that every south pole sticks to a north pole.

When the soap molecules were forced into the water, the hydrocarbon chains necessarily broke apart hydrogen bonds that keep water molecules together as a liquid, and the water molecules tend to become ordered around the hydrocarbon chains. But if the chains gather together, the hydrogen bonds can form again and the ordered water becomes disordered. In other words, the overall entropy increases, and that disordering process outweighs the fact that the soap molecules become more ordered in micelles. This spontaneous ordering of soap molecules happens for the same reason that lipid molecules become organized into the lipid bilayers of membranes that are essential boundary structures of all cellular life.

Self-assembly and encapsulation are the first steps toward life

Everyone has used energy to cause something to happen that could never occur spontaneously in a million years. Imagine that we have dissolved some dish soap in water, put it in a cave, and then came back a year later to check it. Of course, nothing happened. We could have waited a million years, and it would still be sitting there. The reason is that the soap molecules are at equilibrium, floating around in solution as single molecules or in the little aggregates called micelles composed of a few hundred molecules. But now let's add a tiny bit of energy by blowing some air into it through a straw. The amazing result is that soap bubbles form, some floating away in the air. If you had never seen a bubble before, you would be astonished. Simply by adding the energy of a puff of air, the soap molecules have become organized into membranes with a layer of soap molecules on the outside and inside and a layer of water between.

This is called self-assembly, a property of certain molecules like soap. It is a spontaneous process that is one of the foundations of all life today. Every living cell is bounded by a membrane composed of soap-like molecules called lipids, and in the absence of membranes life could not have begun. A biochemist might question whether membranous compartments were essential for life to begin because they can easily set up conditions in which DNA is replicated by enzymes and proteins are synthesized by ribosomes. They could argue that those processes are fundamental to life but don't require membranous compartments. What they forget is that those experiments could not be conducted unless they were confined in compartments called test tubes. It was equally necessary for life to begin in microscopic test tubes formed by self-assembly of soap-like molecules into membranous compartments.

How do we know?

Capturing compounds within membranous compartments composed of lipids might at first seem pretty complicated. If it's so hard, how could it have occurred on the prebiotic Earth? The answer turns out to be remarkably simple. If a mixture of membranous vesicles and a large polymer like DNA is prepared and then exposed to a single cycle of drying and wetting, half of the DNA that was originally outside the vesicles is now inside the vesicles. The reason is that during drying, the vesicles fuse together into a multilayered film in which the DNA is trapped between layers. When the film is rehydrated, the vesicles assemble from the multiple layers but now half of them contain the DNA. This is clearly seen in Plate 14, which shows vesicles containing DNA growing out of a film of dry phospholipid on a microscope slide. The DNA has been stained with a fluorescent dye to make it visible.

The origin of life required a source of energy

Sometimes researchers conduct what is called a thought experiment. For instance, Albert Einstein never did an actual experiment himself but created the theory of relativity by doing experiments in his mind. Let's do a thought experiment to illustrate what we mean by energy. Imagine that we grow some bacteria in a nutrient medium and then use chemical conditions to break down all of their polymers into the monomers that compose them. This is easy to do in the laboratory by heating the bacteria in a solution of sodium hydroxide, commonly known as household lye. The highly alkaline solution causes the bonds linking the polymers to be broken and results in a solution of amino acids, nucleotides (the monomers of nucleic acids), carbohydrates, fatty acids, and phosphate. In other words, everything that composed living bacteria is present in the solution, but it has been broken into small chemical pieces, a sort of scientific Humpty Dumpty. Can a Humpty Dumpty cell be put back together again?

Life requires water, so let's put the chemical pieces into a flask containing freshwater from a volcanic site like the hot springs near Rotorua in New Zealand. Keep in mind that volcanic water is distilled by evaporation from the nearby ocean and then falls as rain, so it is not salty like seawater. Now we wait to see if the pieces of life come back together as living bacteria. If you guessed by intuition that they never will, you guessed right. Nothing would happen no matter how long we wait. Why not? The answer is simple. Making polymers from monomers requires energy, and there is no energy in the flask to make polymers.

So how can we add energy? On the early Earth before life began, there were three sources of energy. Sunlight would be the most abundant, just as it is today, but if we exposed the flask to sunlight, once again nothing would happen because there are no pigments like chlorophyll available to capture the light energy. Another energy source is called chemical energy, but all the chemical energy in the bacteria was lost when they were broken up into smaller molecules. There is one last source of energy, and this is the energy made available when the water in the flask evaporates at the elevated temperatures of volcanic hot springs. The energy necessary to evaporate the water causes the chemical pieces to get more and more concentrated, and when the pieces get completely dry, chemical bonds begin to form between the monomers. For instance, peptide bonds link the amino acids into small chains resembling proteins, and ester bonds begin to link the nucleotides into short nucleic acids. There are also fatty acids in the mixture which assemble into microscopic vesicles if the dry film is rehydrated by rain (see Figure 2.6). The vesicles contain the polymers and are a step back toward the original bacteria, even though they are not alive.

How do we know?

There is abundant evidence that monomers can polymerize simply by drying them into a film at moderately elevated temperatures. Much of the evidence is too technical to describe here, but the most convincing is that we can see the polymeric molecules by a special technique called atomic force microscopy. The images in Figure 2.7 show polymers of RNA synthesized by drying a dilute solution of nucleotides on a sheet of mica. Some of the polymer molecules have formed rings (arrows).

The energy introduced by evaporation is by far the simplest source of energy that would be abundant on the early Earth and would have caused a variety of monomers exposed to wet-dry cycles to polymerize by forming ester bonds and peptide bonds. We can conclude that the polymers required for life to begin were not necessarily invented by life. Instead, the polymers were continuously being synthesized and then encapsulated in self-assembled membrane vesicles to create protocells. Although protocells were an important step toward

DILUTE SOLUTION OF MONOMERS

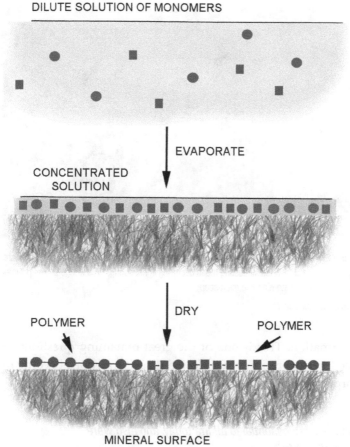

EVAPORATE

CONCENTRATED
SOLUTION

DRY

POLYMER

POLYMER

MINERAL SURFACE

Figure 2.6 Polymers can form when a solution of monomers evaporates. The monomers become extremely concentrated on the surface and then begin to form chemical bonds when completely dry. Candy makers use this process to make taffy, which is a polymer of sugar molecules polymerized by heating and drying.

Credit: Author

the earliest forms of life, the monomer sequences in the encap-sulated polymers were random. In living cells today, the poly-mers contain specific sequences of monomers that are necessary for functions such as enzymatic catalysis or storing genetic

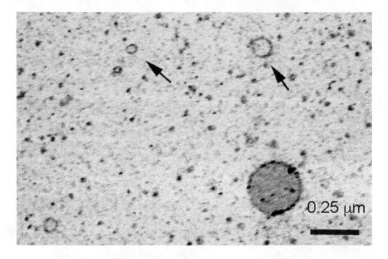

Figure 2.7 Polymers are spontaneously synthesized when the monomers of nucleic acids are exposed to wet-dry cycles simulating those occurring in hot springs. The image shows polymer rings visualized by atomic force microscopy.

Credit: Tue Hassenkam

information. This is one of the great remaining questions related to the origin of life: How did such functions become incorporated into otherwise random-sequence polymers?

Catalysts are essential to all life today, and also were for earliest life

The word *catalyst* in ordinary language conveys the sense that a process is activated by something or someone, but its meaning is different in the language of chemistry. A catalyst in chemistry is something that increases the rate of a reaction but is not itself changed. A common example of an inorganic catalyst is the catalytic converter installed in the engines of cars today. A gasoline or diesel engine combines a liquid hydrocarbon vapor with the oxygen in air to produce a controlled explosion in the cylinders, and the hot gas produced by the explosion drives a piston downward. The downward motion is transferred to rods that cause a crankshaft to rotate and turn the wheels. However, the explosion produces some undesirable products, including carbon monoxide, nitrogen oxides, and unburned hydrocarbons that turn into smog unless they are treated. In this case, the catalyst is a small amount of platinum spread on a ceramic support. As the mixture of potential smog compounds goes through the converter, the catalytic action of platinum transforms them into carbon dioxide, water vapor, and nitrogen gas.

Most enzymes are proteins composed entirely of a hundred or more amino acid molecules linked into a chain by peptide bonds. The sequence of amino acids causes the chain to fold into a precise structure that has an active site where certain amino acids come closely together. Metals are also incorporated into some enzymes. For instance, there are iron atoms in the cytochromes of the electron transport system of mitochondria, and molybdenum is in the enzyme that strips electrons from water during photosynthesis. Perhaps minerals containing metals like iron and copper functioned as catalysts for biologically relevant reactions on the early Earth, and then became incorporated into protein catalysts. For instance, iron oxide, or rust, catalyzes the reaction by which hydrogen peroxide is broken down into water and oxygen. An enzyme called catalase does the same thing in living cells, and iron atoms are part of its active site.

How do we know?

Because thousands of protein enzymes are essential components of all life today, they have been studied by biochemists for a hundred years or more. Their structures have been established by X-ray diffraction, and we know the mechanism of action in exquisite detail. But if you want to stump a knowledgeable enzymologist, ask one simple question: What was the first enzyme? They will say, "Sorry, we don't know the answer yet, but ribozymes composed of RNA could have been the first catalysts used by early life." We will return to the question in Section 3.

Cycling conditions were essential for life to begin

Life today is characterized by cyclic processes, the most obvious being the cycle of growth and reproduction. In the absence of this cycle, life could never have advanced beyond simple chemical reactions. The reason is that life uses genetic information to guide its growth, and in every reproductive cycle small changes called mutations can occur in the DNA sequences that encode genetic information. Most mutations are harmless, but a few are fatal, causing that thread of life to become extinct. Yet another few happen to be beneficial in some way, such as the mutations that allow certain bacteria to resist an otherwise lethal antibiotic. When we look at the fossil record, it is obvious that life has become increasingly complex since it began four billion years ago, and the only way for that to occur is to have cycles of growth and reproduction that allow mutations to accumulate and change populations of organisms.

A less obvious cycle is part of the reproductive cycle of plants. In order to grow, plants must have a source of water, but when plants reproduce they do so by producing seeds or spores that allow genetic information to be scattered into the environment. A seed is mostly dry and survives in a quiescent state in the absence of liquid water. When water returns, the cells in the seed begin to grow and reproduce until they form a mature plant. Wet-dry cycles are a specialty of plants, but a few animals such as tardigrades, or water bears, can also survive drying out. Because cycles are essential for life today, perhaps they were also essential for life to begin.

How do we know?

There is no doubt that wet-dry-wet cycles were abundant on volcanic land masses on the early Earth, just as they are today. The fastest cycles would result from geyser activity: water pumped in large volumes out of the ground into the air falls onto hot rocks surrounding the geyser. The heat causes the spray of water to evaporate within minutes. Longer cycles, measured in hours, are produced by fluctuating hot springs that feed into small pools. As the water level goes up and down, the edges of the pools undergo wet-dry cycles, and in the dry state, films of concentrated solutes remain on the mineral surfaces. The longest wet-dry cycles would be those associated with fluctuations in temperature, producing dew at night that dries during the day, and with precipitation as rain. It is obvious that wet-dry cycles could not occur deep in the ocean but would be common in intertidal zones.

Several physical and chemical properties of wet-dry cycles are important to understand in regard to life's origin. One physical property is related to concentration of organic solutes that was discussed earlier. Recall that dilute solutions of monomers react only slowly or not at all, so it is essential to include a concentration mechanism such as wet-dry cycles. Furthermore, as water evaporates during drying, it increases the energy made available to support synthesis of ester and peptide bonds of biological polymers such as nucleic acids and proteins. However, the polymers have random sequences of monomers. How can they contain genetic information or fold into enzymes having catalytic ability?

Wet-dry cycling can now come into play. The cycles continuously pump energy into systems of molecules that synthesize a variety of encapsulated polymers during drying but then stresses them when protocells assemble in the wet phase of the cycle. Most of the protocells are disrupted and their components are recycled, but a few molecular systems survive the chemical and physical stresses because they happen to contain stabilizing or catalytic polymers. As a result, the initially random polymers increasingly incorporate non-random sequences of monomers and undergo selection for robustness so that stepwise evolution toward the origin of life can begin.

Some chemical reactions increase molecular complexity, others decompose complex molecules

Everyone has heard the old saying that what goes up must come down. Let's think about that for a moment. It takes energy to throw a baseball up into the air because you are doing work against the force of gravity. The energy is stored in the ball and then is released when the ball falls back to Earth. It also requires energy to cause polymerization to occur because water molecules are pulled out from between monomers to make chemical bonds. When a source of energy is absent, water begins to break the bonds in a spontaneous process called hydrolysis, which releases the stored energy. Both synthesis of polymers and hydrolytic breakdown are essential to life. For instance, foods are loaded with polymers like starch and proteins; the only way to get nutritional value from the food is to break the polymers down to their monomers—glucose and amino acids. The hydrolysis that occurs during digestion is catalyzed by enzymes such as amylase and maltase that hydrolyze starch to glucose and proteases that hydrolyze proteins into amino acids.

We described earlier how a simple source of energy can cause monomers like amino acids and nucleotides to be linked by chemical bonds into the polymers required for life to begin. However, those same conditions would have allowed a variety of downhill processes to occur. If polymers are synthesized, they inevitably break down. There must have been a way for them to last long enough for life to begin.

How do we know?

Hydrolysis (from Greek words meaning "water breaking") is the main decomposition reaction in life today. Because hydrolysis is so important, the rates have been measured, and it turns out that peptide and ester bonds are surprisingly stable. For instance, at ordinary temperature and neutral pH ranges, proteins and nucleic acids in aqueous solutions are stable for years and only begin to undergo extensive hydrolysis at elevated temperatures and in extremely acidic or alkaline conditions. Life today is possible because polymers can be synthesized rapidly but hydrolyze slowly. For life to begin, this must have also been true for the first polymers that were synthesized. If the first polymers were hydrolyzed as fast as they were synthesized, life could never have emerged on Earth.

Life depends on cycles of information transfer between nucleic acids and proteins

We can now get down to the crucial question that must be resolved if we are to understand the origin of life: How did the cycle of information transfer in a living cell begin? In life today, this cycle involves enzymes that catalyze transcription of messenger RNA (mRNA) from DNA templates such that the sequence of bases encoding genetic information is transcribed into a sequence of bases in mRNA. The mRNA travels to ribosomes that translate the genetic information into the sequence of amino acids in proteins, and some of the proteins are enzymes that catalyze the synthesis of DNA so that the base sequences are reproduced. This closes the cycle, as shown in Figure 2.8, which illustrates the astonishing complexity of life today.

It seems impossible that such a system could emerge spontaneously on the early Earth, so there must have been primitive versions of the molecules that then slowly evolved increasingly efficient functional capacities. In common usage, the word

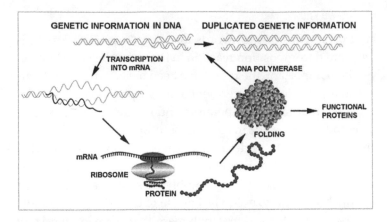

Figure 2.8 The cycle in which genetic information guides the synthesis of catalytic polymers that in turn reproduce the genetic information is a fundamental characteristic of all life today.

primitive tends to mean a simple version of something complex. That sense is appropriate here, but it also has the original meaning of "from the beginning." This leads to a series of questions that need to be answered before we can understand how life can begin on a sterile but habitable planet. The questions will be addressed in Section 3, but here is a partial list:

The arrows in Figure 2.8 imply that something is happening that requires energy. What was the source of energy?

Where did the monomers come from that are required to synthesize nucleic acids and proteins?

How were the monomers polymerized into primitive versions of nucleic acids and proteins?

The sequences of amino acids and nucleotides must have been random at first. How did genetic information become incorporated into nucleic acids, and how did proteins take on the catalytic functions of enzymes?

What was the evolutionary process by which a primitive ribosome emerged?

How was a genetic code established by which a base sequence in DNA was translated into an amino acid sequence?

We understand how the polymer molecules function in living cells today, but reach the end of our understanding when we ask a simple question: How did it all begin?

The oldest known fossil evidence of life is around 3.5 billion years old

The Earth today is very different from the Earth at the time that life began, around four billion years ago. Today the oceans fill about two-thirds of the Earth's surface, and land masses are mostly in the form of continents that float on a red-hot sea of liquid magma. Four billion years ago, small continents had just begun to form and most of the land masses were volcanic. We get a glimpse of what is beneath the continents wherever a volcano breaks through the crust and spews lava. One example is the Kilauea volcano on the Big Island of Hawaii, while others are spread out along the "ring of fire" that marks where the edges of Asia and North and South America meet the Pacific Ocean.

Because of the massive changes that took place during the formation of continents, there is almost nothing left of the original crust of the Earth. Fortunately, a few isolated patches escaped and are still available today for geological research. One of these is called Isua, or in geological terms, the Isua Supracrustal Belt of the North Atlantic Craton in Greenland. Isua is composed of sedimentary rock that was produced when small mineral particles fell to the floor of the ocean. The minerals are mostly composed of limestone and layers of rusty oxidized iron that show up as red bands in the sediment. The age of the rocks ranges from around 3.7 to 3.8 billion years and was established by the rate at which uranium undergoes radioactive decay into lead. No fossil evidence of life has been discovered in Isua rocks. So where have the earliest fossils of life been found?

In the 1980s, Australian geologists began to explore the Pilbara region of Western Australia where there were rocks

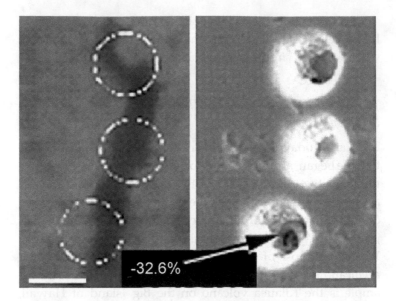

Figure 2.9 This is a micrograph of fossilized bacteria nearly 3.5 billion years old. All that is left of the original bacteria is carbon embedded in silicate mineral. The carbon was analyzed by a complicated method that tells whether it has been processed metabolically. The −32.6% refers to a ratio of two kinds of carbon that exist in nature. One is ordinary carbon with an atomic weight of 12 (six protons and six neutrons in its nucleus) and the second has an atomic weight of 13 because it has an extra neutron. If the carbon was metabolically processed, slightly more of the lighter carbon will be present, and the difference is expressed as parts per thousand. The −32.6 value is definitely lighter than known samples of inorganic carbon, which supports the conclusion that the fossil was once a living organism.

Credit: William Schopf

dated to 3.46 billion years old. They noticed some very strange formations shown in Plate 15 and realized that they were fossil stromatolites produced when bacteria formed mineralized layers as they grew. In fact, living stromatolites are still growing in Shark Bay, just 200 miles away. Other researchers collected ancient rocks from the same region and began to look for actual fossils of bacteria that might have been preserved in the sediments. The results became very controversial when some scientists suggested that the blurry microscopic particles were not actually fossils, but there is now a consensus that at least some of them are real (Figure 2.9).

Section 3

WHAT WE STILL NEED TO DISCOVER

The title of this book is *The Origins of Life: What Everyone Needs to Know*. Well, everyone needs to know not just the answers we have found so far but also the questions that remain. This is because the cutting edge of science lies not in the answers but rather in all the unanswered questions. Scientists who are working on the questions are guided by their ideas, yet very often these ideas don't agree with each other. That doesn't mean the conflicting ideas should be discarded. Instead, we evaluate the experimental or observational evidence and then decide how much explanatory power they have.

Some of the most significant remaining questions are described in this section. If you are a student thinking about getting into a scientific career, you could spend the rest of your life investigating any of the following questions. That's how important they are.

Is the RNA World real, or just conjecture?

When someone thinks up an idea about how life might have begun on Earth, they often give it a name to make it easy to remember. Examples include the Iron-Sulfur World, the Lipid World, and the RNA World. The last of these worlds was suggested by some of the most prominent scientists thinking about the origin of life, including Francis Crick, Carl Woese, and Leslie Orgel, and has become a paradigm for several reasons. Probably the most important was the realization that certain kinds of ribonucleic acids are catalysts: because they are composed of RNA rather than proteins, as enzymes are, they are called ribozymes. The idea that RNA can serve as a catalyst and also store genetic information prompted Walter Gilbert to propose that the first forms of life did not begin with the complex interactions of DNA, RNA, and proteins, but instead relied solely on RNA in an RNA World.

A good hypothesis states a question and possible answers; it also makes predictions. For instance, if the first life used only RNA, we should find remnants of that in today's life. In fact, quite a few have turned up. Compounds based on ribonucleotides play essential roles in metabolism, for example ATP, the energy currency of all living cells. More astonishing, the active site of protein synthesis in ribosomes is a ribozyme, hinting that the first forms of life actually did use RNA both as a catalyst and to store information.

Although these observations support the RNA World hypothesis, substantial gaps in our knowledge remain to be filled. We don't know yet how nucleotides could be synthesized on the early Earth nor how they could be polymerized into RNA molecules long enough to serve as ribozymes. We do know that biological RNA is prone to hydrolytic decomposition. So even if it can be synthesized, how could it last long enough to take part in the processes leading to life in an RNA world?

These are significant concerns, but no one works on the origin of life without being endlessly optimistic. Time and again seemingly insoluble problems have turned out to have surprisingly simple answers. A hundred years ago, when Oparin and Haldane first began to think about how life could begin, no one would have imagined that amino acids could be available on the prebiotic Earth. That's why the result of Stanley Miller's spark experiment was such a revelation, followed closely by the discovery that the Murchison meteorite contained no less than seventy organic compounds classified as amino acids.

Nucleobases have also been detected in the Murchison meteorite, again confirming that an essential component of life can be synthesized by naturally occurring chemical reactions. Adenine is one of the four nucleobases present in nucleic acids. No one would have guessed that it might also have been available on the early Earth, but Juan Oro showed it can easily be produced from hydrogen cyanide (HCN).

To summarize, the concept of an RNA World has had immense value as a working hypothesis to be tested experimentally. The current aim is to discover a way to fabricate a simple system of encapsulated molecules in the laboratory that uses catalytic RNA (ribozymes) to grow and reproduce. The ultimate test will be to see if the system can function in natural conditions such as the hot springs that would have been common on the prebiotic Earth.

What is metabolism and how did it begin?

Metabolism is the network of enzyme-catalyzed reactions that transform organic molecules into products necessary for maintaining life. Thousands of biochemists have worked for a hundred years to understand metabolism and as a result we know the individual steps in remarkable detail. But do we really understand how metabolism could have been incorporated in the earliest forms of life?

Five main processes are involved:

1. Nutrients are transported into cells and changed into compounds that support growth.
2. The energy content of nutrients is captured and utilized.
3. The energy and nutrients are used to synthesize polymers that function as enzymes and structural components.
4. Other polymers are synthesized that store and utilize genetic information.
5. Damaged polymers are broken down and turned over, a process called catabolism.

The easiest way to understand the first process is through an example. Glucose is a source of nutritional energy and is processed in all living cells by a metabolic pathway called glycolysis. This is one of the simplest metabolic systems, yet it involves ten distinct reactions, each catalyzed by an enzyme. How could such a complex series of reactions emerge from the chaotic mixture of organic compounds and energy sources on the prebiotic Earth? Of course, we don't know yet, but we do know that if chemical energy is available, mixtures of compounds will react to form more complex molecules. Formaldehyde (HCHO), for example, can react with itself to form compounds such as ribose and glucose. We also know that formaldehyde can react with hydrogen cyanide to produce amino acids. These chemical reactions are not used by life today, but they do illustrate the possibility that a primitive version of metabolism could have emerged from spontaneous reactions occurring in small ponds in which organic compounds had accumulated.

We are just scratching the surface of what might be possible. For instance, phosphate is essential to all life today, but we only know a few reactions by which phosphate can be added to organic compounds in solution. Yet this is exactly what happens in the first step of glycolysis. If we can discover not just the reaction, but also a plausible catalyst, we might begin to understand how glycolysis, one of the most important metabolic reactions, could have been incorporated into the earliest forms of life.

What were the first catalysts?

In life today, it is clear that RNA in the form of ribozymes can act as a catalyst, a discovery that gave rise to the idea of an RNA World. But ribozymes are not very efficient catalysts. There is a biochemical term called turnover number, which is defined as the rate at which reactant molecules are converted into a product by one molecule of the catalyst. A really fast ribozyme molecule might have a turnover number of fifteen reactions catalyzed per second, but protein enzymes have turnovers thousands of time faster. One of the fastest is an enzyme called catalase, which has a turnover of ten million per second.

Even though ribozymes might have been the first catalysts, most biological catalysts today are protein enzymes composed of hundreds of amino acids linked together in a very precise order guided by genetic sequences in DNA. Such complicated molecules could not have popped up spontaneously on the early Earth, so there must have been simpler substances that served as catalysts. There are many possibilities, some of which involve the chemical properties of certain metals, and a good example is the catalase mentioned earlier. In order to understand why catalase is important, we need to know something about hydrogen peroxide, which is a water molecule with one extra oxygen: HOOH instead of HOH (H_2O). Most electrons in chemical compounds come in pairs, so HOOH can be written as HO:OH in which the two dots represent an electron pair that forms the chemical bond linking the two OH groups. But peroxide has a tendency to break the bond spontaneously: HOOH → 2 HO•. The little dot after HO represents an unpaired electron, and compounds with an unpaired electron are called radicals. They are extremely reactive and can damage polymers like DNA and proteins. This is why hydrogen peroxide can be used as a disinfectant.

We know that hydrogen peroxide is produced by mitochondria as a byproduct of oxidative metabolism, so why isn't it toxic in the cell? The reason is that cells are protected by catalase. If a pinch of catalase is added to a bottle of 3 percent hydrogen peroxide, bubbles suddenly appear and foam may even flow over the top. The catalase has accelerated the rate of the reaction by breaking the hydrogen peroxide down into water and oxygen. The scientific definition of a catalyst is something that makes a chemical reaction proceed faster toward equilibrium without itself being changed and catalase fits that definition. It made the reaction go a million times faster but remains unchanged after the reaction was over.

So what is in catalase that causes the reaction to speed up? Analysis of catalase showed it consisted of four subunits composed of amino acids. Each subunit has an iron atom bound within a specialized molecule called porphyrin. The iron porphyrin complex by itself can also act as a catalyst breaking down peroxide, although not nearly as fast as catalase. In fact, we can even add iron oxide, or rust, to hydrogen peroxide, and all by itself iron will cause the reaction to proceed faster.

The point is that very simple compounds, particularly metals, can act as catalysts. Iron atoms in catalase cause it to have a distinctive green color, and the iron in blood gives hemoglobin its distinctive red color. Iron is also in the cytochrome enzymes of the electron transport chain of mitochondria that transfers electrons to oxygen from a source like pyruvic acid produced by glycolysis. Cytochromes are also red, and the word *cytochrome* in fact means "cell color." Both iron and copper atoms are in cytochrome oxidase that catalyzes the final step in which electrons are delivered to oxygen, and the green color of plants is due to a magnesium atom in the center of the chlorophyll molecule that lets it capture light energy. The electrons released from chlorophyll are the source of energy for all life on Earth today. We don't know yet what the first catalysts were in the earliest forms of life, but some of them were likely to incorporate atoms of metals like iron, copper, and magnesium.

How did regulatory feedback loops begin to function?

Regulatory feedback loops are often overlooked when we think about how life can begin, probably because there is a natural tendency to work on simpler reactions such as synthesis of biomolecules like amino acids and nucleobases. Feedback loops become important when we go to the next level in which systems of polymers begin to function in concert.

To understand what is meant by feedback loops, consider what would happen if there were no regulatory feedback between the air temperature and the furnace in your home. You would either be much too hot or much too cold, and that is why a thermostat is necessary to control the temperature. If the air in the house is too cold, the thermostat senses the cool temperature and turns on the furnace, which puts out heat and raises the temperature of the air. When the air reaches the desired temperature, the thermostat turns off the furnace to close the loop. Life is the same. In the absence of feedback control, the processes of life would be chaotic! Metabolic reactions would be too fast or too slow, and too many or too few of life's primary components would be synthesized. For this reason there is an intricate system of regulatory feedback loops in place in all life today.

Let's consider a few biological examples, some of which are familiar, others hidden at molecular levels of organization. At the level of the organism, our bodies have a thermostat embedded in the nervous system of the brain. If we are too cold, we shiver to generate heat, and if we are too hot, we sweat to get rid of extra heat by evaporation.

At the physiological level, if glucose concentration in the blood is too high, the pancreas secretes more insulin so that the glucose can be transported more rapidly into the cells of muscle and adipose tissue. If there is too little glucose in circulation, another hormone causes glucose to be released from glycogen stored in the liver.

At the molecular level, there needs to be a way to control the rate at which a catalyst is active. If there is too much of its product, it must be slowed down, and if too little it should speed up. For instance, ATP contains chemical energy in the bond between the second and third phosphate in the triphosphate linked to the main molecule. That bond is hydrolyzed to produce adenosine diphosphate (ADP) and phosphate when energy is needed. Now suppose that you decide to go out for some jogging. When you exercise, the energy stored in ATP is used to drive muscle contraction with the result that ADP and phosphate accumulate in the cell and the amount of ATP decreases. Because ATP inhibits four of the enzymes involved in metabolic energy production, when the level of ATP falls, those enzymes can speed up and so more ATP is synthesized.

How could the first forms of life develop such regulatory feedback loops? This fundamental question has rarely been addressed, but a pioneering paper published by Aaron Engelhart and Kate Adamala in collaboration with Jack Szostak demonstrated a simple feedback mechanism. In the experiment, a ribozyme composed of two strands of RNA was used. The ribozyme could act on itself by breaking one of the chemical bonds holding it together, but only if both pieces were together. The ribozyme was encapsulated in vesicles along with a high concentration of small oligonucleotides that bind to the ribozyme and keep the two pieces from coming together and being activated. Then more fatty acids were added to simulate growth of a cell. The vesicle membranes grew, and this diluted the internal concentration of the inhibiting RNA which fell off and allowed the two strands of the ribozyme to come together and become active.

What a complicated experiment! And yet, this is one of the simplest examples of feedback, and it is not even a feedback loop because the signal goes in only one direction when dilution activates the ribozyme activity. If it were a true feedback loop, the ribozyme would make more of the shorter inhibiting RNA and turn itself off when the concentration got too high. But this example does serve to illustrate the difficulty of understanding how feedback loops began to regulate metabolic functions in the first forms of cellular life.

How did life become homochiral?

There is a deep mystery about life on Earth that no one has been able to resolve yet. But before revealing the mystery, we need to know the meaning of three words: *chiral, racemic,* and *enantiomer.*

The word *chiral* comes from a Greek word having to do with hands and has been incorporated into several English words. For instance, *chiropractic* is a way to treat a variety of health problems by using hands, and *Chiroptera*, meaning "hand-wing," is the scientific name for bats. So chiral must have something to do with hands.

An obvious property of hands is what we call handedness, meaning that we have right and left hands that cannot be superimposed. In other words, a right-handed glove doesn't fit on your left hand and vice versa. Some organic molecules, but not all, also have a handedness. Carbon has four chemical bonds when it is part of a molecular structure, and the bonds are usually in a tetrahedral arrangement with the carbon atom at the center. If all four bonds are attached to different chemical groups, the compound will be chiral, but not if two or more of the groups are the same.

To illustrate this point, we can consider two amino acids called glycine and alanine. The carbon atom of glycine is chemically bonded to two hydrogens, one nitrogen in an amine group ($-NH_2$) and another carbon in a carboxyl group ($-COOH$), while the carbon atom in alanine is chemically bonded to one hydrogen, one nitrogen, one carboxyl carbon, and one methyl group ($-CH_3$). In other words, all four of the atoms or groups attached to the central carbon of alanine are different, and this is what makes it chiral while glycine, with two hydrogen atoms on the central carbon, is achiral. Another way to think about this is that the mirror images of glycine can be superimposed, but the mirror images of alanine cannot be (Figure 3.1). The mirror images of chiral molecules are called enantiomers, and a mixture of two enantiomers is said to be racemic.

The study of chirality might simply be another branch of chemistry except for one fact: All life uses just one of the two possible enantiomers of amino acids and carbohydrates. In other words, life is homochiral. Amino acids (except glycine) are L-enantiomers and carbohydrates are D-enantiomers. (L is

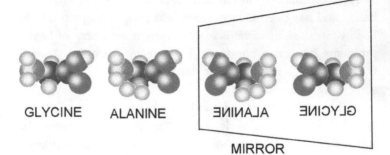

GLYCINE   ALANINE   ∃NINA⅃A   ∃NIƆYⅬG

MIRROR

Figure 3.1 The structures of glycine and alanine are shown as reflections in a mirror. Turn the mirror image of glycine 180 degrees and you will see that it superimposes perfectly on the original image. Do the same with alanine and it cannot be superimposed, just as your right hand won't fit into a left-handed glove. That property makes it a chiral molecule, while glycine is nonchiral.

Credit: Author

the abbreviation of *levo-* from the Latin word for "left," and D is an abbreviation of *dextro-*, meaning "right.") Although we don't yet know how life became homochiral, we do know why homochirality is essential. Think of the structures of life as analogous to a giant jigsaw puzzle with all the pieces fitting together perfectly. In other words, they are homochiral. But then make the pieces racemic by turning half of them upside down. They can never fit together into a complete puzzle. The amino acids and carbohydrates of life's polymers are like the puzzle pieces. They can only fit together in a polymer if they are homochiral.

The organic compounds available on the prebiotic Earth were synthesized by chemical reactions, not biological enzymes, so they must have been racemic. How did the first forms of life fit them together into functional polymers? There are lots of ideas, but no consensus yet. One of the simplest is that the first polymers related to life were synthesized non-enzymatically and were therefore composed of racemic monomers. As they went through cycles of synthesis and hydrolysis, any polymers that happened to have an excess of L or D monomers were more stable or perhaps more efficient as catalysts. If so, they would be selected and quickly become dominant in the race to the first forms of life.

What is photosynthesis, and how did it begin?

We go about our daily lives, breathing oxygen, eating cereal and milk for breakfast, and enjoying the lawn and trees we can see through the window. We never stop to think that all this is possible only because some microbial cells began to capture light energy around four billion years ago. Their discovery has been a continuous fount of biological energy ever since and literally changed a sterile early Earth into a planet where the human race can thrive. No one yet knows how photosynthesis began, but we do know how to find the answer. We can begin by understanding how photosynthesis works today and then think of a way that a primitive version could have captured light energy four billion years ago, perhaps even before life began.

What really happens when sunlight illuminates a plant? We learn in high school science classes that something called chlorophyll is responsible for the green color of plants. What is usually skipped over is that the green color is not important. It's just light that remains after the chlorophyll captures the red and blue light that was absorbed. Plate 16 shows the structures called chloroplasts in plant cells where all the reactions of photosynthesis occur. When researchers first isolated chloroplasts from plant tissue, they were surprised to discover that they contained DNA with base sequences related to those of cyanobacteria. This means that plants contain organelles descended from the same bacterial species that first evolved photosynthesis and generated the oxygen that is now the dominant source of energy for animal life.

What does the chlorophyll do with the light it absorbs? This is where it gets complicated, but the basic principle is not too hard to understand. The electronic structure of chlorophyll captures the photons of light and jumps from a ground state to an excited state. This is analogous to a tuning fork that can generate a sound of 440 vibrations per second, which is middle A on a piano. If you listen to the fork in a quiet room, you don't hear anything. But now take the fork out into a city street where it is exposed to a mixture of sounds, what we call white noise. If you hold the tuning fork up to your ear, you will hear it vibrating and producing the tone of middle A. Some of all that white noise is at 440 vibrations per second. The tuning fork absorbs exactly that energy, a process called resonance, and then releases it as the tone you hear. Chlorophyll is like a tuning fork, but one that absorbs only the red and blue light in the white sunlight that hits it. While the tuning fork responds to a frequency of 440 vibrations per second, red and blue light have frequencies of a trillion waves per second!

Now we can describe how the first step of photosynthesis works. The excited state of chlorophyll doesn't just vibrate when it captures the energy of a photon. Instead, it gets rid of the extra energy by releasing an electron into a set of proteins called an electron transport chain, at the end of which the electrons end up on a compound called NADP (nicotinamide adenine dinucleotide phosphate). It's astonishing to realize that those electrons represent the energy source for virtually all life on Earth because they are used to change carbon dioxide into carbohydrates. Furthermore, as they travel down the electron transport chain, their transport is coupled to a reaction that builds up a proton gradient across the membrane. The energy of the gradient is used to synthesize ATP, which drives the metabolic reactions that turn carbon dioxide into carbohydrates. And finally, the electrons given off by chlorophyll must be replaced. Where do they come from? Water! A very high-energy reaction in chloroplasts strips electrons from water, leaving behind the oxygen that is used as an energy source by animal life, including humans.

How did such a complicated process begin? It certainly didn't involve chlorophyll, a very complex molecule that can only be synthesized by enzymes. There must have been simpler molecules that were abundant and could absorb light energy so as to be promoted to an excited state that could donate electrons. One possibility is the polycyclic aromatic hydrocarbons (PAH) that are probably the most abundant organic compounds in the universe. They are present in carbonaceous meteorites so we know that they can be synthesized by non-biological chemistry and delivered to the Earth's surface. Furthermore, pioneering work has shown that their excited state can donate electrons to other molecules and also fix carbon dioxide. Future research may reveal ways that the PAHs became embedded in membranes and perhaps served as the beginning of photosynthesis.

What was the first ribosome?

This is one of the deepest questions related to the origin of life. Ribosomes are incredibly complex molecular machines that translate genetic information from DNA into nearly perfect sequences of amino acids in proteins. The simplest ribosomes are those of bacteria, and early studies showed that they were composed of one large and one small subunit. Eukaryotic ribosomes have the same basic structure with a few additions.

How was the structure of ribosomes established? They are much too small to be seen with an ordinary microscope that uses light, so X-rays are used instead. If a solution of table salt is allowed to evaporate slowly, beautiful cubic crystals begin to appear in the solution. A century ago, an experiment was done in which a beam of X-rays was aimed at one such crystal placed in front of photographic film. The amazing result was that the beam produced a pattern of dots on the film. This is called a diffraction pattern and is caused by the X-rays being reflected off layers of organized atoms within the crystal. The atomic structure of the crystal could be deduced from the pattern.

A hundred years later, several laboratories succeeded in making crystals of ribosomes and then used X-ray diffraction to deduce how the atoms were arranged in the molecule (Plate 17). This required years of work because a single ribosome contains 140,000 atoms! The leaders of the three teams that determined the crystal structure were each awarded a Nobel Prize.

Now that we know the structure of a ribosome we can begin to think about how primitive ribosomes might have become an essential component of early life. Loren Williams and his research group at Georgia Tech have made some progress by dissecting the RNA and proteins of ribosomes and deducing which portions seem to be the most ancient. The scheme illustrated in Plate 17 shows their progress so far. They must assume that there was an unknown source of the small RNA molecules shown in Phase 1. Step by step the ancestral RNA bound other forms of RNA to become increasingly complex, finally adding proteins in Phase 6 to produce the actual ancestor of the large subunit.

This is elegant, pioneering work but still leaves questions open: Where did the RNA molecules come from in the first place, and at what point did the ribosome begin to use encoded base sequences in mRNA to synthesize proteins? There is still much to learn about the origin of ribosomes in the earliest forms of life.

How did the genetic code emerge?

When the double helix structure of DNA was reported in 1953 by James Watson and Francis Crick, it was a revelation because the structure explained how genetic information could be stored, used, and replicated from one generation to the next. Little by little, the puzzle pieces of life began to be fit together by generations of scientists. The next question was to determine how the information was stored and how it was used by the ribosome to guide protein synthesis. The process that Francis Crick called the Central Dogma turned out to involve storage of information in the base sequences of DNA, transcription of the information into mRNA, and translation of sequence information into the amino acid sequences of proteins by ribosomes.

Plate 18 shows a ribosome in action, with the large subunit on top and the small subunit below. MRNA is moving through the active site from right to left, and a transfer RNA (tRNA) is carrying an amino acid that will be added to the growing B chain of an insulin molecule.

Like any other kind of coded information, the genetic code translates base sequences in the mRNA into amino acid sequences in a protein. A simple analogy is Morse code in which letters symbolized by dots and dashes are transmitted electrically over wires from a sender to a receiver who then translates the patterns of dots and dashes back into letters. In the 1960s, the problem was to understand how base sequences in DNA and RNA could encode the information required to guide the sequences of amino acids in proteins. Obviously, one base per amino acid could not do the job because there were four bases and twenty amino acids. Two bases per amino acid can be grouped into sixteen combinations, which is still not enough. But three bases per amino acid were more than enough because the four bases taken in groups of three could be arranged in sixty-four ways. We now know that base triplets called codons carry the genetic code, and that most amino acids are specified by several different triplets. There are also start and stop codons that tell a ribosome where to begin and end translation from the mRNA into a protein.

Although we can deduce a little evolutionary history by comparing the base sequences in ribosomal RNA from bacteria with those of more complex organisms, there is still no experimental evidence that tells us how such a complicated process became incorporated in the first forms of life.

Where did viruses come from?

Viruses are everywhere in the biosphere. In fact, it has been estimated that there is as much viral biomass in the ocean as there is biomass of living organisms. Virtually everyone has personal experience with viruses associated with the common cold and influenza, and new viruses such as HIV and Ebola make headline news when they trigger epidemics. What we don't see is the constant warfare between bacteria and the viruses called bacteriophages ("virus eaters") that is going on all around us.

If viruses are so abundant today, were they also abundant when life began? Could viruses have been the first form of reproducing molecules that later gave rise to cellular life? A few scientists think so, but a more common opinion is that they are a kind of molecular parasite that evolved the ability to reproduce by using the protein synthesis apparatus of living cells. Viroids are even simpler than viruses and were discovered because they infect and damage potatoes, avocados, peaches, and coconuts. They turned out to be circular RNA molecules having just 246 to 467 nucleotides. The image in Figure 3.2 shows a viroid and for comparison a virus that causes the common cold. Typical viruses are composed of both proteins and nucleic acids and reproduce in living cells by taking over the cells' protein synthesis machinery to make copies of themselves. Viroids have a much simpler life cycle. After getting into a plant cell, they are reproduced by an enzyme called RNA polymerase that normally synthesizes mRNA. The enzyme reels off hundreds of copies of the viroid, which then begin to interfere with other protein synthesis processes.

Viroids were discovered in 1971 by Theodor Diener, who proposed in 1989 that they may be a kind of molecular fossil left over from an RNA World. Although viroids today require living cells to reproduce, we can speculate that simple versions of nucleic acids might have assembled spontaneously on the early Earth and then began to reproduce by non-enzymatic

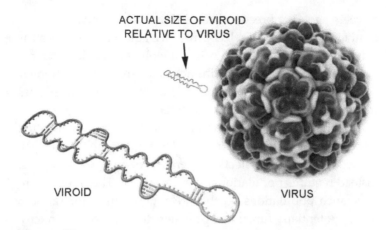

ACTUAL SIZE OF VIROID
RELATIVE TO VIRUS

VIROID

VIRUS

Figure 3.2 Viroids are the smallest known infective agent and are composed of little more than a few hundred nucleotides in a ring-like structure of RNA. They are much smaller than a typical virus particle, as shown in this image.

Credit: Prepared by author from public domain images

catalysis. At some point they became encapsulated in membranous compartments as the first step toward the RNA World.

How did encapsulated polymer systems begin to evolve?

The word *evolution* comes from a Latin word meaning "unrolling" and is now defined generally as changes occurring over time, such as stellar evolution of stars going through their life cycle. Charles Darwin applied the word to the changes that occur in populations of living organisms and give rise to different species. Now, 150 years later, we know that the changes are the result of mutations in DNA that lead to differences in individual organisms. These produce variations, within populations, which can undergo selection. Darwin called this "natural selection" and used artificial selection of domestic animals as an example. For instance, wolves began to interact with humans thousands of years ago, and early humans began to select wolves that exhibited friendly behavior. Hundreds

of years ago, dog breeders realized that selection was a powerful tool and began to select physical characteristics as well as
behavioral. As a result, the original wolves have descendants
ranging in size from tiny chihuahuas to Great Danes. Similar
agricultural selections have produced corn and wheat species
from plants that resemble weeds today.

We can now ask a fundamental question: what were the first
steps of biological evolution? We don't know the answer yet,
but we do know that selection operates not on individuals, but
instead requires populations with variations. The only way to
have such populations on the prebiotic Earth is for random
sets of potentially functional polymers to be present as microscopic protocells. Simple arithmetic shows that a milligram of
amphiphilic lipid compounds can produce a trillion protocells
comprised of vesicles with encapsulated polymers. Plate 19
shows how protocells can emerge when a film of dry lipid is
exposed to water. Polymers that have formed during the dry
phase of a wet-dry cycle become encapsulated within microscopic protocells. At first, each protocell is different in composition from all the rest. A natural version of combinatorial
chemistry then becomes possible in which protocells are like
microscopic test tubes in a natural experiment. Most are inert
and will be disrupted and their components recycled, but a
rare few will happen to have properties that let them survive
intact into the next cycle. This represents the first step of selection in an evolutionary process. When the dry phase of a wet-
dry cycle begins, the membranes of surviving protocells will
fuse and deliver their polymer cargoes back into the multilayered lipid matrix to participate in another round of synthesis.

What are the properties of successful protocells? The properties emerge from a synergy between the components of the
system: polymers and the surrounding membrane. All of the
processes up to this point have been driven by self-assembly
and a very simple source of energy—the chemical potential
of dehydration that drives ester and peptide bond synthesis
so that monomers can form polymers. Now, however, rare

protocells may have encapsulated systems of polymers with specific functional properties such as vesicle stabilization, pore formation, metabolism, catalyzed polymerization, and replication. We do not yet know which polymers have these properties, but a reasonable guess is that they resembled RNA and peptides.

What are progenotes and LUCA, the last universal common ancestor?

Unless you are a biologist thinking about the earliest forms of life, the word *progenote* is probably new to you. It was coined by Carl Woese and George Fox to describe a hypothetical stage of early evolution in which the relationship between genotype and phenotype had not yet been fixed. In bacteria today, as cell growth and division occur, the next generation of cells is virtually identical to their parents in terms of genetic composition and gene products. But before the first true bacteria appeared on Earth, it seems inescapable that there were systems of encapsulated molecules that had a primitive metabolism and were capable of growth and some form of reproduction. Woese and Fox predicted that in the transition through the progenote phase toward life, an immense amount of genetic information would have been shared as populations of early protocells competed for efficient ways to maintain specific sets of polymers.

We have speculated that such processes were enabled within wet-dry cycles, and my colleague Bruce Damer proposed that populations of protocells went through a moist, aggregated "gel" phase on the way to complete dryness (Plate 20). In the gel phase, the protocells no longer maintained individuality but instead interacted and fused so that their encapsulated polymers were mixed and shared. In an evaporating pool, these aggregates of protocells would be exposed to increasing concentrations of dissolved solutes. If some of these were potential nutrients, they would be available for primitive metabolic processes.

Given protocells and progenotes, we can speculate how they might evolve increasingly complex molecular systems that approached the transition to what we call life. An important concept to keep in mind is that the first steps toward life occurred in mixtures of organic compounds exposed to energy sources. If the conditions favored polymerization of amino acids into peptides, the same conditions would favor polymerization of nucleotides into nucleic acids. Instead of assuming that one of the processes was "first," it is better to think that polymer systems were co-evolving from the start. Primitive ribosomes are an example. The first ribosomes required both RNA and peptides, so an obvious conclusion is that both polymers must have been available.

Over millions of years of cycles and natural experiments, progenotes discovered and shared interacting systems of polymers that increased their ability to survive environmental stresses. When stable aggregates of progenotes emerged, other experiments could begin. These are obvious, and a list of functioning polymers includes catalysts of primitive metabolism, catalysts for polymerization reactions, nucleic acids to store and transmit genetic information, ribosomes to translate genetic information into functional proteins, pigment systems to capture light energy, electron transport systems to generate proton gradients, and a way to couple proton gradients to ATP synthesis. As other forms of energy became available, the energy of wet-dry cycles would no longer be necessary. A system of polymers that enabled protocells to divide into daughter cells is the final step toward allowing them to capture other forms of energy.

Finally, after perhaps half a billion years of natural experiments the first forms of life emerged. These are referred to as LUCAs or the last universal common ancestors, because they had all of the functional molecular systems that are still present in life today. With a genetic code common to all of life they coalesced into the trunk of the tree of life.

How did prokaryotic life become eukaryotic life?

When biologists first began to use a microscope to study living cells, they noticed that some cells had a nucleus, but bacteria did not. The nucleus looked like a little nut, or kernel, so cells with nuclei were called eukaryotes from Greek words meaning "good kernel" and bacteria were called prokaryotes, meaning "before a kernel" assuming that simpler prokaryotic life came first during early evolution. That assumption was correct, and for two billion years after life began the only living organisms on Earth were prokaryotes, mostly photosynthetic cyanobacteria thriving in lakes and the ocean. Furthermore, the photosynthesis was oxygenic because the cyanobacteria had evolved mechanisms not just to capture light energy but also to use that energy to pull electrons away from water. The electrons were used to change carbon dioxide into the carbohydrates required for metabolism and growth. But when electrons are pulled out of water molecules, what is left over is molecular oxygen.

For millions of years, the oxygen was used up in a very simple reaction which is familiar to everyone: rusting. The seawater in the ocean had lots of iron in solution in the form of iron ions with two positive charges, abbreviated Fe^{++}, or ferrous iron. When ferrous iron ions collide in solution with oxygen molecules, or O_2, the oxygen grabs another electron from the iron to make it ferric iron, or Fe^{+++}. Moreover, the oxygen atoms form chemical bonds with the ferric iron to produce a compound that contains two iron atoms and three oxygen atoms, abbreviated Fe_2O_3. This is insoluble in water and precipitated onto the ocean floor, forming layers composed of a mineral called magnetite. The result was immense deposits of what we now call iron ore.

The iron in the ocean used up all the oxygen produced by cyanobacteria for a billion years, but finally no more ferrous iron was left. At that point, the oxygen began to enter the atmosphere, and about two billion years ago this led to the GOE, the Great Oxidation Event. We can see this in the geological record of ancient soils called paleosols, which abruptly change from a grey color to red when iron in the soil began to turn into rust. The abundance of atmospheric oxygen was the first step toward life as we know it today because energy was made available when electrons stripped from water by photosynthesis were given a pathway to return to molecular oxygen. It allowed a new form of microbial life to develop that did not depend on light energy because an even greater energy source was made available by oxygen in the atmosphere. Finally, after two billion years, life could begin to experiment with greater complexity.

The natural experiments could proceed along several alternative pathways, one of which was that two different single-celled organisms could combine into a new form of life, a process called symbiosis. When this was first proposed by Lynn Margulis in 1973, other scientists were very skeptical of the idea. How could such symbiosis happen? Over the next ten years techniques for analyzing nucleic acids improved and it became clear that the DNA in prokaryotes was in the form of a circle containing several million nucleotides while in eukaryotic cells the DNA was mostly in the nucleus. Out of curiosity, some scientists began to ask whether DNA might be present outside the nucleus, and the answer was astonishing. Circular molecules of DNA were present in mitochondria and chloroplasts! Furthermore, when the base sequences were established, mitochondrial DNA sequences matched those of alphaproteobacteria and chloroplast DNA sequences matched cyanobacteria.

This was a revelation: All advanced forms of multicellular life on the planet—plants and animals—depend on organelles descended from bacteria to make energy available. Each of our cells contains hundreds of mitochondria that grow and divide when the cells divide, and their DNA has been carried along ever since the original symbiotic combination nearly two billion years ago.

How the symbiosis actually occurred is an open question, but one possibility is simply that a large bacterial cell somehow "ate" a smaller bacterium, but instead of being digested the small bacterium set up housekeeping and thrived as a kind of parasite. We know that such processes are still occurring today. One of the best experimental examples is a single-celled species of amoeba that can engulf toxic bacteria as a food item. Most of the amoebae die from the toxin but a few survive, and the bacteria not only live in the amoebae but begin to provide energy by photosynthesis. Over many generations the amoebae begin to depend on the energy. This can be demonstrated by adding an antibiotic to the culture medium that will kill the internalized bacteria. The amoebae, now lacking a source of energy, will also die.

Is there a Tree of Life?

When Darwin first conceived of the theory of natural selection and evolution, he made a little sketch in his notebook (Figure 3.3). Then, with characteristic caution, he added a little note off to the side: "I think."

In 1879 Ernst Haeckel decided to illustrate Darwin's concept and drew an actual tree with names of various kinds of animal life hanging off the branches. Single-celled amoebas were at the bottom, and gorillas, orangutans, and humans were at the top. The simplified idea of a tree of life was easier to understand than Darwin's lengthy arguments, and Haeckel's image helped fix the idea of evolution in the minds of British and European thinkers.

The first suggestion that a tree might not be the best way to think about the history of life came from Carl Woese's brilliant idea that evolutionary history should be reflected in sequences of ribosomal RNA. His reasoning is easy to understand. Given that there is a slow accumulation of mutations over time, there will be relatively few differences within the RNA of bacteria,

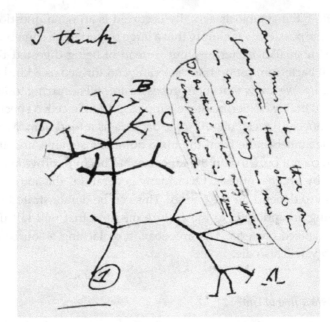

Figure 3.3 Charles Darwin's tree.
Credit: Adapted from Wikimedia Commons

and if that is used as a reference, there will be increasing numbers of mutations as we compare the bacterial RNA to that of ever more complex species. When Woese examined the RNA of various microorganisms, he had a huge surprise. As well as prokaryotes and eukaryotes, he discovered a new microorganism that he decided to call archaea. So, after life's origin, LUCA gave rise not to a tree of life but to three branches now called kingdoms: Archaea, Bacteria, and Eukarya.

Later research showed that not only were there three distinct kinds of life, but that genetic information was being shared among progenotes, a process called horizontal gene transfer. Furthermore, the mitochondria and chloroplasts of eukaryotes actually contained their own DNA, which could be traced back to ancestral prokaryotes called alpha proteobacteria and cyanobacteria. This could not possibly fit into a tree of life! In his article "Uprooting the Tree of Life," published in *Scientific American* in 2000, Ford Doolittle proposed a shrub as a better metaphor for life's history than a tree with branches emanating from a central trunk; in particular, a shrub in which organisms continuously exchange and combine genetic information, as they have right from the beginning.

Over the past twenty years, DNA and RNA sequencing has vastly expanded our knowledge of evolution, so much so that even a shrub proved an inadequate metaphor. The history of life is now represented as a circle with the origin at the center and a vast amount of evolutionary branching toward the periphery (Figure 3.4). It's humbling to see the evolutionary path of *Homo sapiens* not as a route to the top of a tree but instead just one more of the countless paths that led to life as we know it today.

Can we synthesize life in the laboratory?

Richard Feynman was a brilliant physicist who won a Nobel Prize for his discoveries in nuclear physics. He was also a superb speaker who made every effort to help his students understand abstruse physical concepts. During one of his talks, Feynman wrote on a blackboard "What I cannot create, I do not understand."

It does seem possible that we now know enough about biochemistry and molecular biology to be able to create a simple version of life in the laboratory. No one has done it yet, but we can at least think about how cells might be taken apart and then put back together again. The ultimate aim of a new subdiscipline of biology called synthetic biology is to do just that. If we think about the parts list of life, something that seems impossible becomes probable.

Here's a parts list for a bacterial cell, each with a specific property that is essential for life:

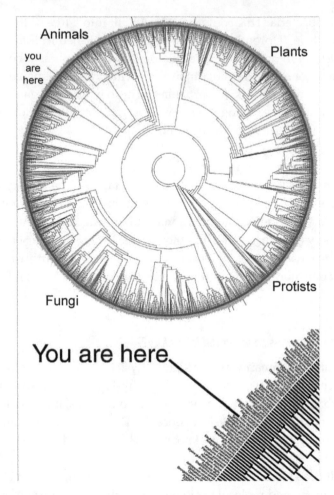

Figure 3.4 Because living organisms continuously exchange and share genetic information, there is no tree of life. Instead, the evolutionary history of life on Earth is viewed as a vast spreading circle in which the origin of archaea and bacteria is shown at the center. These remain intact all the way to the outer perimeter. However, near the center you will see a point where prokaryotic microorganisms combine symbiotically to form eukaryotic life that undergoes further evolution into protists (protozoa), plants, animals, and fungi. The "fuzz" around the outside represents vast numbers of named species, including *Homo sapiens*, shown in the lower, right-hand corner in a magnified view.

Credit: Adapted from Global Genome Initiative

1. Lipid membranes spontaneously assemble into microscopic compartments required for cellular life. As described in Section 2, it's easy to encapsulate large molecules in the compartments simply by drying a mixture of the lipid.
2. Ribosomes can be easily isolated. They are stable, and for many years researchers have mixed ribosomes with mRNA for a specific protein and synthesized that protein by a process called translation.
3. Isolating a circular bacterial genome is harder because they tend to break. However, with care it can be done. In fact, researchers at the Craig Venter Institute have already synthesized an entire genome of a small bacterial species from scratch and then put it back into a cell whose DNA had been deactivated. The bacteria began to grow even though they were using genetic information from an entirely synthetic genome composed of DNA.
4. There are thousands of enzymes in a typical bacterial cell. No one is going to synthesize those from scratch, so we need to use the ones that are available and have already been synthesized by living bacteria.

The bottom line is that ALL of the essential parts of bacterial cells have been shown to work in isolation. However, no one has ever tried to put them back together. Is this even possible? Can a mixture of bacterial parts that is not alive be revived? Let's propose a thought experiment.

We know how to use an enzyme called lysozyme to dissolve the cell walls of certain kinds of bacteria, leaving a little membranous bag called a protoplast that contains all the components of a living cell. We also know that the bags can be popped open by putting them into water because they swell up and burst, releasing their components as shown in Figure 3.5.

Now comes the trick we will use to put the protoplasts back together. We will prepare lipid vesicles from lipids that were extracted from the bacteria and add them to the mixture

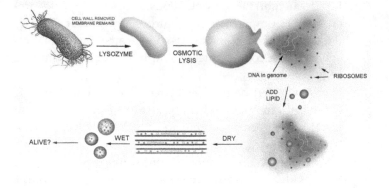

Figure 3.5 How to reconstruct a living cell.
Credit: Author

of functional polymers that was released when we popped open the protoplasts. The last step is to let the water evaporate under vacuum in a refrigerator. As the lipid vesicles become increasingly concentrated during drying, they will fuse into thousands of lipid layers, and all of the bacterial components will be crowded together between the layers. When we add a dilute solution of nutrients, the layers swell and capture the components into trillions of microscopic compartments.

Are they alive? Will they grow and reproduce? After all, the ribosomes, genomes, and enzymes are all back together in one place. Most knowledgeable scientists would say, "No! They will NOT be alive!" But they can't know for sure, because no one has done the experiment. I tend to share their skepticism—for a very good reason. All the components of the cell may have been put back together in a tiny membranous bag, but we have disrupted an invisible order having to do with feedback loops that regulate metabolism. In the absence of feedback controlling thousands of enzymes, it may be impossible for the cells to come back to life.

Nonetheless, pioneering scientists have actually tried something like this. Albert Libchaber and Vincent Noireaux at

Rockefeller University and Tetsua Yomo in Japan extracted the intracellular components from bacteria and encapsulated them in lipid vesicles along with DNA containing the gene for green fluorescent protein (GFP). The Rockefeller scientists also included a second gene for a protein called hemolysin that made the lipid membranes permeable to amino acids and ATP. When they provided a nutrient solution to the artificial cells, they began to glow with a green fluorescence, which meant that the entire protein synthesis pathway was working and GFP was being synthesized.

This does not mean that the cells are alive, only that one protein was being synthesized. The next obvious step is to repeat the experiment to see how many of the 5000 genes in the bacterial DNA are being translated into functional proteins. This might seem impossible, but I hope someone will find out for sure whether the mixture of encapsulated components can assemble into a simple version of life. We might be surprised.

Could life begin again on the Earth today?

If Charles Darwin had been asked whether life could begin on Earth today, he would have said, "Probably not!" He hinted at this in his famous note to Joseph Hooker in 1871: "at the present such matter would be instantly devoured, or absorbed, which would not have been the case before living creatures were formed."

The point Darwin was making is that compounds required for life to begin are in fact nutrients, and microbial life today is so efficient at using nutrients that even if a primitive form of life somehow got started it would immediately be gobbled up.

But there is another problem, which is the presence of oxygen in today's atmosphere. We tend to think of oxygen as life-giving, but that's because we have evolved ways to use it as a source of energy to drive metabolism. The energy is made available by stripping hydrogen from food and letting the electrons cascade down the electron transport chain to oxygen. We can only do that because we also have multiple ways to protect cell components from toxic effects of oxygen. For instance, vitamin E is one of the protective antioxidants that works by inhibiting the propagation of oxidative damage in membrane lipids. If it is left out of mouse diets, in a month or two their health begins to deteriorate. They become anemic and lose mobility because their blood cells are damaged by all the oxygen that flows through cell membranes as it is transported around the body.

Oxygen can also degrade many compounds that would otherwise be nutrients. This effect can be seen in the surface of freshly cut apples that quickly turn brown, or in bruised bananas. The unpleasant rancidity of oils and wines is also a result of oxidation damage.

The bottom line is that oxygen is so reactive that organic compounds would not last long enough to take part in the reactions leading to the origin of life. This would not have been a problem in the atmosphere of the prebiotic Earth, which was mostly composed of unreactive nitrogen gas together with small amounts of carbon dioxide. Because photosynthetic oxygen was virtually absent on the prebiotic Earth, organic compounds could circulate in solution long enough to support the origin of life.

Could conditions on other planets allow life to begin?

This question is a driver for NASA and ESA (European Space Agency) scientists who are exploring other planetary bodies in our solar system, and also for astronomers studying exoplanets in orbit around other stars in our galaxy. Humans are endlessly fascinated by the possibility that the Earth may not be the only place where life emerged on a sterile yet habitable planet. NASA has managed to land three rovers, Spirit, Opportunity, and Curiosity, on the surface of Mars. In 2020 another advanced rover will begin its voyage to Mars with the explicit goal of searching for evidence that microbial life

once existed there. It is even possible that intelligent life exists elsewhere and may have developed sufficiently high technology to broadcast radio signals in an attempt to communicate. Radio astronomers realized that sensitive antennas might be able to detect such signals, and this gave rise to multiple projects falling under the rubric of Search for Extraterrestrial Intelligence (SETI). In either case, the largest hurdle is simply whether conditions on other planets would allow life to begin and evolve into increasingly complex forms.

Based on the information presented in this book, is it possible for life to originate elsewhere? One way to approach this question is to consider where life exists on our planet, but also where it cannot exist even after three billion years of evolution. This puts constraints on the kinds of conditions conducive for life and incorporates our knowledge of extremophilic life. The main constraints involve availability of liquid water, temperature, pH, and concentration of common ions. Let's consider those one at a time.

At the Earth's surface, water is a liquid between 0°C and 100°C. Below 0°C, water is solid ice and above 100°C it is a vapor. By observation, living organisms can be preserved in solid ice but cannot grow and reproduce because the necessary nutrients and energy cannot diffuse to support metabolism. For instance, liquid water does not exist in the high desert regions of Antarctica nor in the Atacama Desert in Chile, and neither place supports microbial life capable of growing and reproducing.

At the other end of the temperature scale, volcanic hot springs have water temperatures approaching the boiling point of water, between 90°C and 100°C depending on their altitude. Deep in the ocean the pressure is so great that liquid water can exist at even higher temperatures, and certain bacteria can survive a temperature of 121°C. We can conclude from the temperature extremes that life could not begin on a planet that has water in the form of ice or a desert world that has land but no liquid water. Mars is an interesting test of this conclusion.

Its surface today is much harsher than the Atacama Desert in Chile, but ice exists at its poles and beneath the surface. There are no active volcanoes today, but an immense volcano the size of France was apparently erupting 100 million years ago. We also know that 3.5 billion years ago Mars had shallow seas over some of its surface. Given the facts described in Section 2, we can conclude that life could have begun on Mars when it had active volcanoes and hydrothermal hot springs. A reasonable prediction is that someday Mars rovers will detect evidence of early microbial life resembling what we see today in the fossil stromatolites of Western Australia.

Will we ever know how life can begin?

The answer is: Maybe.

There are many ideas and approaches to the question but no consensus. The ideas are being evaluated by a jury of scientists who judge them in terms of explanatory power and the weight of evidence. I am a member of the jury, so a good way to end this book is to present an alternative hypothesis for the origin of life that can be tested by future research. The components of the idea were described earlier and can be easily summarized:

Life began in hot spring water that was distilled from a salty ocean and fell on volcanic land masses.

Organic compounds delivered by meteoritic infall and geo-chemical synthesis accumulated in the hot spring water.

Some of the organic compounds were monomers that poly-merized, while others were amphiphilic molecules that spontaneously assembled into membranous structures.

The hot springs experienced continuous cycles of wet and dry conditions due to fluctuations in water levels and evaporation.

During drying, the mixture of organic monomers and amphiphilic compounds became extremely concentrated,

and polymers were synthesized by condensation
reactions.

Upon rehydration, the polymers were encapsulated in
membranous vesicles to form vast numbers of protocells.

Each protocell was different in composition from all the
others in terms of its content of polymers with random
sequences of monomers.

Everything up to this point has been verified experimentally
in the laboratory or by observation of how water behaves in
hot springs associated with volcanic activity. The next steps are
speculative but can guide future research. The ideas emerged
from a collaboration with my colleague Bruce Damer, and they
are illustrated in Plate 21.

Most protocells within a population were inert and their
components were recycled, but a rare few contained polymers and membranes that exhibited properties relevant
to life processes. Examples of such properties include stability, selective permeability, and catalytic activity.

Protocells with such properties survived the endless cycling
and slowly began to dominate the population. This was
the first step of Darwinian evolution.

During time intervals measured in millions of years, the catalytic polymers became incorporated into systems that carried
out primitive metabolism with regulatory feedback, captured
chemical and light energy, and used the energy to catalyze
their own reproduction from nutrients available in the environment. Those systems crossed the threshold from non-living
protocells to the first forms of cellular life. Again over millions
of years, the populations migrated downhill toward the ocean,
slowly adapting to increasingly salty water and finally be-
coming sufficiently robust to thrive in intertidal zones. These
microbial populations formed the mineralized stromatolites

that survive today in Western Australia as fossil evidence for the earliest known life.

I hope that this book has given readers a sense of the excitement of scientific research. The ideas expressed may ultimately be shown to have explanatory value as they undergo further testing, or they may need to be discarded if they fail. This is the process by which we will finally understand how life can begin on the Earth and other habitable planets.

Further reading

I decided not to include references to the scientific literature in this book because the language used is often highly technical and the papers are not readily available. In fact, most journals require payment for access. However, it seems worthwhile to point to several books published in 2018 and 2019 that were written in non-technical language and provide additional information for interested readers.

My own book is called *Assembling Life* (Oxford University Press, 2019) and was written to describe a new hypothesis that freshwater hot springs undergoing wet-dry cycles are conducive to the assembly of molecular systems. An open-access journal article by Bruce Damer and myself describes the hypothesis in detail (The Hot Spring Hypothesis for an Origin of Life. *Astrobiology*, April 2020).

Without realizing it, over the past two years Stuart Kauffman and I were both writing books that were published by Oxford University Press in 2019. Stuart is the author of several other books, including *The Origins of Order* (1993), *At Home in the Universe* (1996), *Investigations* (2002), and *Humanity in a Creative Universe* (2016). In his latest book, *A World Beyond Physics: The Emergence and Evolution of Life*, Stuart proposes that the origin of life may go beyond the known laws of physics. Physicists might disagree, but will be stumped if they are asked which laws could predict that a puff of air causes a soap solution to assemble into beautiful membranous bubbles. A soap bubble

can be understood in hindsight, but no physical laws would have predicted its existence. This is referred to as an emergent phenomenon, and the same question can be asked about the origin of life. The laws of physics and chemistry may someday allow us to predict how non-living matter can become alive, but not yet.

Robert Hazen is the Director of the Deep Carbon Observatory in which a large group of scientists investigated the physical, chemical, and biological distribution of the element carbon on Earth. Out of this experience, Hazen wrote *Symphony in C: Carbon & the Evolution of (Almost) Everything* published by W.W. Norton & Company in June 2019. Besides his work as a scientist, Hazen is a concert musician as a trumpet soloist, and he combined his knowledge of mineralogy and music to write a unique perspective on the role that carbon plays in our lives.

Dirk Schulze-Makuch and his co-author Louis Irwin published the third edition of *Life in the Universe: Expectations and Constraints* (Springer) in 2018. A title like this is pretty bold because we have no idea whether there is any life in the universe except on our own planet. However, the authors use their knowledge of chemistry, physics, astronomy, and biology to convincingly argue that there is a high probability that life exists elsewhere, even though it might not look like life on Earth.

Bahram Mobasher is a professor at the University of California, Riverside. He wrote *Origins: The Story of the Beginning of Everything* (Cognella Academic Publishing, 2018), which is designed to be used as a textbook in a course he teaches. The book lives up to its title. It really is about everything from the origin of the universe to the origin of life.

Quite a few earlier books shaped my own perspective on the origin of life and are worth reading. Here is a short list of recommended titles:

Pier Luigi Luisi. *The Emergence of Life: From Chemical Origins to Synthetic Biology.* 2nd ed. Cambridge University Press, 2016.

Nick Lane. *The Vital Question: Energy, Evolution and the Emergence of Complex Life.* W.W. Norton & Company, 2015.

Eric Smith and Harold J. Morowitz. *The Origin and Nature of Life on Earth: The Emergence of the Fourth Geosphere.* Cambridge University Press, 2016.

Peter Ward and Joe Kirschvink. *A New History of Life.* Bloomsbury Press, 2015.

Addy Pross. *What Is Life?* Oxford University Press, 2012.

Freeman Dyson. *Origins of Life.* Cambridge University Press, 1999.

INDEX

Figures are indicated by *f* following the page number

For the benefit of digital users, indexed terms that span two pages (e.g., 52–53) may, on occasion, appear on only one of those pages.